$16.50

Inter-relationships of the Evolutionary Systems

By Bahman K. Shahrokh

For some thirty years, Dr. Shahrokh has studied the problems of the origins of life and matter—and this book presents, for the first time, the fresh and persuasive answers he found.

During his dedicated research he soon became convinced that the origins of life, the Earth, the solar system, stars and matter are closely inter-related. Then his studies gradually took him into sciences not directly related to biology, as he explored deep into the evolutions of matter and life. Now, in this remarkable volume, Dr. Shahrokh's hypotheses inter-relate the diverse major evolutionary systems in the universe. He writes:

"There are eleven chapters in this book. The first one gives a concise summary of my criticisms of the present theories and the manner of my approach in finding alternative explanations. In the second chapter some of the fundamental scientific principles related to the topics under consideration are discussed. Chapter III elaborates on the contents of the second chapter with the intent of showing how distance can cause redshift; it also lays the foundation for the hypothesis describing the creation of matter and the origin of universe in the fourth chapter. The rest of the book devotes itself to the use of the deductions arrived at in the first four chapters in explaining the origins, the evolutions and the inter-relationships of the major evolutionary systems. In the final chapter, a new theory is proposed regarding the origin and evolution of life. The book concludes with a discussion of the probable role of Man in the interwoven processes of the evolution of matter and life."

Inter-relationships of the Evolutionary Systems will be a needed addition to the bookshelves of the scientific world.

Inter-relationships
of the
Evolutionary Systems

Inter-relationships
of the
Evolutionary Systems

Bahman K. Shahrokh

Since 1891

Binford & Mort

Thomas Binford, Publisher

2536 S.E. Eleventh • Portland, Oregon 97202

Inter-relationships of the
Evolutionary Systems

Printed in the United States of America

First Edition 1977

PREFACE

I have spent a good part of my last thirty years in attempting to understand the reasons for the existence of certain relationships among the evolutionary systems in the universe. The project began in 1946 when I developed an interest in the problems of the origin of life. It grew into a major undertaking after I became convinced that the origins of life, the Earth, the solar system and matter are closely inter-related. This in turn brought about a prolonged involvement in fields not directly related to biology. I believe the endeavor has been worthwhile, as it has produced a series of hypotheses tying the seemingly diverse evolutions together.

The theories in this book are founded on the concept that the properties of matter are the result of specific patterns of motions of "quanta in their corpuscular state." When the motions are haphazard and disorganized, we cannot comprehend them; but when the corpuscles move relative to each other in orderly patterns, they become recognizable as forces, energies and subatomic particles. Forces and energies are then regarded as the different manifestations of the same phenomenon and components of a continuous spectrum ranging from gamma rays to nuclear forces.

In these explanations, forces are produced by the spins of subatomic particles, and electromagnetic waves by their oscillations. The corpuscles that are transformed to force move uniformly away from the subatomic particles with the speed of light; those that are converted to electromagnetic waves propagate in the same manner, but intermittently, and consequently they acquire the wave properties. These hypotheses are then used to describe how distance and gravitational forces cause redshifts. These concepts, however, do not negate the theory of the expanding universe; they merely suggest rates of recession slower than the present accepted values.

The theory of creation of matter follows these discussions. It is based on the concept of the transformation of the disorganized motions of corpuscles in the state of "nothingness" to the organized motions of subatomic particles, forces and energies. This conversion takes place through the process of "pair production," a self-perpetuation phenomenon that is still in progress at the periphery of the material universe. In this proposed process, equal numbers of *hydrogen* and *antihydrogen* atoms are produced. The mode of distribution of these atoms and their interactions then determine the manner of the formation of stars and galaxies. The origins and evolutionary paths of the solar system, stars, nebulae, galaxies and matter are discussed in successive chapters, the hypotheses being based on the concept of generation of energies by stars through *matter-antimatter reactions.*

The final chapter contains a new theory regarding the origin of life, based on the pattern of the Darwinian evolution. Here, it is postulated that the pre-cellular evolution set the pattern from which the post-cellular evolution became a replica, both following the pattern of evolutions of stars and galaxies.

In the preparation of the contents of this manuscript I have received considerable help from three persons: Ruth M. Chesbro, my associate of many years who, in the face of considerable doubts on my part, devotedly encouraged me to proceed with the work; my wife, Agnete, who patiently went through my writings and gave me many useful suggestions as to how to put my thoughts in readable language; and, finally, my son, Peter, who edited the manuscript and assisted me in making the book ready for publication.

B.K.S.
Berkeley, California
Bend, Oregon

CONTENTS

LIST OF ILLUSTRATIONS

xiii

Inter-relationships
of the
Evolutionary Systems

I

THE PROBLEMS

As long as men have been thinking, they have pondered about the origin, existence and development of life and the universe. Throughout the ages, their search for the answers to these questions has led them to accumulate immense quantities of scientific information on subjects ranging from the remotest galaxies to the most intimate structures of atoms and living cells. But despite this quantity of information, they have yet to formulate an explanation which can tie up the diverse components of their knowledge into one satisfying whole.

To accumulate knowledge and enrich their understandings, scientists have depended largely on experimentations, observations and the interpretation of the observed results. But in studying the evolutionary trends, the investigators have been severely handicapped by the fact that most of the major evolutions have not left enough traces of their origins to permit the drawing of reasonable and satisfying conclusions. Thus with the scarcity of clues, reliance has been placed on controlled conjectures and logical but arbitrary suppositions to guide the patterns of thoughts at times when no empirical reality presented itself. Yet, for the most part, the currently accepted theories regarding the origin of the universe and its various components cannot be regarded as the real answers. They are based on highly speculative assumptions, lack continuity between the considered past and present situations and fail to include the interrelationships of the different evolutionary systems. In addition, some cannot even be supported by the known scientific principles.

1

An example of such incongruity is found in the theory of the expanding universe, which has been derived from the dual interpretations of the redshift observed in the spectra of distant galaxies. On one hand the redshift is considered to be a function of distance and is used to determine the galaxies' distances from the Earth; the greater the shift in the observed spectra the greater the distance. On the other, by the application of the Doppler effect's principle, the velocities of the recession of distant galaxies have been deduced to be also functions of their redshifts; the greater the shift the greater the rate of recession. Here, two unrelated values, the distance from the Earth and the rate of recession, have been obtained with one observation of the spectral shift; this makes one wonder if correct interpretations have been derived from these observations.

The relationship between a luminous object's distance from its point of observation and its redshift is a straightforward phenomenon infringing on no other scientific principles and determined by repeated observations. Yet as an index of the velocity of a source's recession, the redshift is also an indisputable phenomenon when dealing with nearby stars and galaxies. But it possesses two drawbacks in its applicability to the field of cosmological investigation: first, its application produces calculated values for the rates of recession of some of the distant galaxies falling into the unlikely range of the relativistic velocities; and second, the phenomenon of the redshift is not caused solely by the Doppler effect—it is also observed when light passes through an intense gravitational field. The reason for not accepting distance as the true cause of the redshift of lights coming from the distant galaxies is the lack of any scientific explanations as to why distance should cause the shift. Yet this does not mean that this effect does not exist or that no explanation can be found.

The interpretation of redshifts in terms of both distance and recessional velocities initiated the belief that there has been one point of origin containing all of the matter in the universe from where the universe began to expand. Such a belief then led to the inevitable conclusion that the universe originated with an immense explosion. But to explain this conclusion, many compromises were made with scientific principles. In order to appreciate these violations, one merely has to try to explain logically the structure of the original mass before the explosion, its origin, the cause and the mechanics of the explosion and the changes that took place in the mass after the fantastic blow-up. The inconsistencies and improbabilities are readily apparent. Furthermore, how can one account for the transformation of the supposedly extreme dense matter of the primordial mass to the basic units of the universe, the hydrogen atoms?

The theory of the origin of new stars, based on the condensation of gases and dust particles, also veers away from the fundamental principles. Ingenious attempts have been made to reconcile how atoms of gas and particles of dust, supposed products of the disintegration of degenerate stars, join together again to form gigantic masses of almost pure hydrogen and then turn into new stars by the effects of the gravitational forces. But these attempts have overlooked a simple and basic premise: matter turns gaseous because the repulsive radiation energies emitted from its atoms are greater than the attractive gravitational forces emanated from the same number of atoms. This is why gases expand rather than contract in unconfined space.

As far as the author is aware, no experimental evidence yet shows that masses of atoms and molecules of hydrogen have a tendency to come together under the conditions of extremely low pressures and comparatively high radiation, the conditions existing in interstellar space. The exception, of course, is at temperatures near 0° K, a condition not normal within galaxies. The claim that the accretion of dust particles causes condensation does not stand close scrutiny either; if solid particles of dust are responsible for the formation of these masses, then young stars should be composed either of degenerate atoms or of the high atomic weight elements usually present in the dust particles. This assumption directly contradicts the origin theory for the new stars, which is dependent on the presence of almost pure hydrogen for the start of the fusion reactions.

In the conventional theory of the stellar origins, the relating of how the fusion of the hydrogen atoms begins is generally avoided. One cannot positively deduce that sufficient pressure is developed at the centers of the supposedly highly compressed masses to raise the temperatures to the required level of atomic fusion. Certainly in a sphere of gas, an uneven distribution of the cumulative effects of these forces exists. But can one say confidently that there are almost no limits on the quantities of forces that can accumulate in one point, especially when one knows that the pressure caused by these gravitational forces produces correspondingly stronger repulsing energies? And even if such pressures do start the fusion of hydrogen, no convincing explanation yet tells how the nuclear reactions in the center of the intensely compressed and immensely explosive hydrogen mass are contained. The argument stating that the physical expansion of gases in such a huge structure will have a controlling effect on the rate of the explosive reaction is unsubstantiated. Yet this precarious mechanism is supposed to have caused the formation of countless stars in the universe.

There is nothing new in these criticisms; the uncertainties may be noticed in the manner these theories are presented in the literature. The general attitude toward these explanations is one of hesitant acceptance; their faults are known but they are the best presently available, and it is believed that they can be modified slowly as new facts are received. But though this encourages the pursuit of new information, the approach is basically flawed. It accepts the fundamental assumptions as correct and believes all that is needed are modifications of the formulated concepts which will be determined by the accumulation of new data. An obvious but more difficult alternative may be necessary: begin from the beginning and take a new approach in interpreting the already available facts.

Because the answers to the old questions are to be found within the major evolutionary systems, it seems sensible to commence the search for a new interpretation with a study of the inter-relationships between these systems' characteristics. This is the reasoning: as the evolution of life is a product of the evolution of the solar system, itself an outgrowth of the evolution of the Galaxy and hence the universe, repetitions of certain fundamental patterns in these systems should be expected. The best example of such a common pattern is that of a single strong central force surrounded by a number of sources of weaker forces, all functioning to form a whole system in equilibrium.

The structure of the basic unit of matter, the atom, follows this pattern. Many molecules are built in the same way. The solar system operates on the same principle, and almost all galaxies show the same pattern. Other examples are found in cells, the principal units of living organisms; here nuclei are the dominating centers. Even in the social patterns of human society, from the family to the most complex systems of government, this pattern is copied. The repetition of such patterns among the evolutionary systems may be termed "the phenomenon of pattern replication." This phenomenon may then be used to advantage in understanding and explaining certain aspects of evolutionary systems.

Because the evolutionary trends are inter-related, one should expect the general pattern of their beginnings, their periods of existence within changing states of equilibrium, and their ends to resemble each other. Furthermore, the comparison of the life spans of these systems should yield certain conclusions not possible to obtain in any other way. For instance, say nothing of importance is known about a certain period in a specific system's evolutionary trend, but information is available for its equivalent in another evolutionary system. This information may then be used to formulate what occurred in the first system during the unknown period in question. Thus an explanation of the evolution of matter can be developed by comparing it to the evolution of life. The latter has a definite, well-known trend; the slow, persistent process of its beginning resulted in the building and organization of the living cells. Once established, these gave rise to their own kinds of entities through their use of the material in their environment. A diversification in body structures and functions then followed and resulted in a variety of new living units which established a state of equilibrium among themselves. The trend can be predicted to continue for some time; after a while it will come to an end, if only because the Earth, the seat of known life, is not going to last forever.

Now compare this knowledge and prediction with the evolutionary trend of matter. The beginning of the universe is still unexplained. There exists a considerable amount of information about the intermediate phase where an equilibrium has been established between the diversified units of matter, and one knows that the known universe is moving toward oblivion just by observing the way matter is being converted to energy in all the stars and dissipated into space.

If the similarity between the evolutionary patterns of life and matter is acknowledged, one may conclude that the beginning of the evolution of matter was also a slow, persistent process where units of matter formed new units of matter by causing a duplication of their structures. Thus initially from a state of "nothingness" one of each of the four fundamental and permanently stable elementary particles (proton, antiproton, electron and positron) came into being. From then on these particles catalyzed the formation of similar particles from the same medium in the same pattern as cells produced cells in the beginning of life.

Though this type of reasoning was of help, it was too general in scope to produce acceptable interpretations of the results of the observations already at hand. Two major problems had to be faced: (1) an explanation that could be checked by experiments and observations for distance causing redshift had to be provided; and (2) a substitute hypothesis supported by the known scientific principles had to be formulated to explain the origin of the universe. The answers to the two problems did not need to be totally inter-related. If the redshift could be satisfactorily explained in terms of distance, then the problem of the origin would be simplified to the extent that there would be no need to use a gigantic explosion as the mainstay of a theory.

The solution to the problem of the redshift lies in the understanding of the true nature of the electromagnetic waves. The question of the creation of matter is deeply rooted in the structure of matter. Both are related to the fundamentals which connect energy, force and mass together. My attempts in explaining the two problems will be based on the use of reasoning while staying within the realm of reality as much as possible. What follows then is a discussion of these newly suggested solutions for the age-old questions. Hopefully it will be experiments and observations carried out in response to these proposed hypotheses, not the use of unsubstantiated theories, which will act as the final judges for the verification or refutation of these interpretations.

II

RE-EXAMINING THE BASICS OF MOTION, FORCE AND ENERGY

Motion

"Motion," defined as the state of change of distances between the existing units, may be said to be the cause of all the observed phenomena in the universe. In this respect four important facts stand out: (1) motion is a universal phenomenon and as such, with the exception of some motions on Earth, is beyond Man's control; (2) all phenomena in the universe are different manifestations of motion; (3) no event ever occurs without the participation of a number of motions; (4) the perception, the interpretation, and the understanding of the events require constant evaluation of the rates of motions, each of which is the result of a large number of simultaneously occurring motions. To summarize: *the universe may be described in terms of motions of countless units with respect to each other, the size of the units ranging from the largest stars to the smallest known units, quanta in their corpuscular state; this makes the observation of any phenomenon in this perpetually moving and ever-changing limitless expanse the determination of the properties of a set of localized interacting motions which have established or are in the process of establishing a state of equilibrium.*

The ability to perceive events, a faculty highly developed among birds and mammals, has been inherited by Man from his primitive ancestors. The interpretation and the understanding of events, however, is a recent acquisition which began to take place when Man gained the ability to associate the events occurring before him with those that occurred previously. In this association he had to compare the rates of motions in the process of taking place with those which had already occurred and had left their impressions in his brain. From these he made attempts to foretell the future. Man calls this very complex inter-relationship between the rates of real motions and those produced by his mental faculties "time," and he has used it to great advantage to understand many of the events that automatically take place in nature.

Motion and Time

We may start the discussion of the concept of "time" with two hypothetical but pertinent questions: (1) if Man did not possess the faculty of perceiving and the means to measure mentally the rates of motions, would he have conceived the phenomenon of "time"?; and (2) if Man did not know anything about "time," would the phenomenon still be there?

The answer to the first question is simple. It can be argued that without the capability of estimating the rates of motions, the perception of motions would not have been an orderly and organized affair; this would have prevented Man from recognizing the lengths of intervals between events, and the concept of time would not have been born. The second question, however, is not easy to answer, but the following discussion will try to provide a reasonable solution.

One of the fundamental factors connected with the phenomenon of "time" is the role of the physiological functions of the human body in the perception and estimation of the rates of motions. The physical and mental activities of Man are entirely dependent on the regular movements of molecules, atoms and subatomic particles within his body; thus the judgment of the rates of the observable motions is closely related to the rates of the physical and chemical reactions taking place inside his brain. In most cases the central nervous system automatically takes care of the perception and estimation of the rates without a person even being aware of the process. This is clearly illustrated in the following examples:

1. On a freeway one continually estimates the rates of movements of the vehicles nearby.

2. An athlete can easily coordinate his own movements with that of a ball moving toward him; he simultaneously estimates the rates of two sets of motions: those of his body and of the ball.

3. From the rate of the vibration of his tympanic membrane, a trained person can name a note played on a piano.

These are only simple examples selected from our daily activities. But they point to the fact that without the ability of using our inborn time mechanisms for the estimation of rates of motions, we would be totally helpless in anything we undertake.

When there is a need for the more accurate measurement of motions than the brain can provide, we resort to the use of instruments where the rates of motions of objects under investigation are compared to those of an arbitrary standard. The usual practice is to measure the length of the interval between two recognizable events in terms of a standard periodic motion; this measurement consists of counting the number of times the standard periodic motion passes a relatively stationary point during the interval. The standard can be the revolution of the Earth around the Sun, the rotation of the Earth around its axis, the swinging of the pendulum of a clock, the regular change of directions of an alternating current, the frequency of a monochromatic electromagnetic wave emitted by oscillating electrons, or any other appropriate periodic motion. But no matter what standard is used, the basic principle for the measurement of "time" is always the same. Because of history and tradition, the revolution of the Earth around the Sun and the rotation of the Earth around its axis are the two most widely accepted standards. All other periodic motions used for the measurement of time are calibrated in terms of these arbitrary standards.

But though Man uses instruments for the more or less precise determinations of the rates of motions, his principal way of estimating the rates of motion is by the direct use of his central nervous system. It is common knowledge that the process of life is composed of a large number of interwoven physical and chemical cycles where each cycle, similar to any periodic motion, follows a cyclical pathway and constantly repeats itself. These metabolic cycles are used in the nerve cells as standards to judge the rates of the incoming impulses. An impulse may be described as a locally developed point of electrochemical potential produced in response to a stimulation at the end of a neuron process, traveling toward the main cell body along the cell membrane. On reaching the main nerve cell, the potential leaves its impression on the cytoplasm, a complex mixture of compounds whose function is controlled by the clockwork operation of the metabolic cycles. Since the incoming series of impulses are produced in response to the effects of the external motions on the receptors, the method of estimating the rate of an external motion by a nerve cell is through the comparison of the rates of the arrivals of the impulses with the rates of the metabolic cycles. *Therefore, the faculty of estimating the lengths of intervals between successive events without the use of instruments is the process of comparing the rates of motions that are taking place outside the body with the rates of the biological cycles that are taking place within the body.*

The power of evaluating the rates of motions is a significant part of the evolution of all animals because the survival of any species is, to a large degree, dependent on the capabilities of its members to judge the rates of motions of the prey or the enemy. This faculty of estimating the rates of motions is just as important to the primitive cells as to the more advanced animals. For example, in the living pattern of a *Paramecium* cell, the rhythmic motions of the cilia, the recognition and ingestion of a food particle and the digestion and the excretion of the by-products all require precise timing by the components of a single cell. Naturally, the problem of survival also includes the question of the probabilities of encountering food, adversaries, and other contingencies of life. But unless the motions of the components of the bodies are thoroughly coordinated and controlled automatically through internal timing mechanisms, no advantage can be taken of the opportunities that come along.

One reason Man gained supremacy in the animal kingdom was because he learned to forecast the future from his knowledge of the past. With the evolution of his brain this art has been refined to such an extent that most of our present-day endeavors and hopes for success depend on correctly forecasting the future from the experiences of the past. This approach has caused the acceptance of the idea that there is substance to both the "past" and the "future." This, of course, is far from the truth, the reason being that the mechanisms behind the knowledge of the past and the foretelling of the future are the processes that are taking place at "present." For example, when we read history, we are using the written words as signals; we imagine we are experiencing the past, whereas in reality the entire vision is caused by chemical and physical cycles of the body that are taking place at "present." This "present" nature of each thought is evidenced by the fact that if the man who is engrossed in the events of the past accidentally breathes enough hydrocyanic acid gas, his visions and images of the "past" disappear. It happens because the toxic gas interferes with one single step in his metabolic cycles. We are not implying here that the events of the past did not occur, but rather that the past does not exist and that what reminds us of the "past" are events taking place at present.

The difference between the "past" and the "future" lies in the fact that the former has already taken place and has left referential relics and other evidence behind, while the latter is the product of the conclusions that we arrive at by our constant use of the law of probability. For example, when we say that there will be a tomorrow, we actually mean to say that from the experiences of the past we are concluding that the Sun will most probably rise once more. *Hence, the "present" is composed of the summation of all the motions occurring at the instant (an extremely short interval during which our senses or our instruments fail to register the extent of the motions) that they are being perceived; the "past" signifies all the movements that have already taken place and which can be recalled by using the mechanisms operating at "present"; the future consists of the expected motions which may or may not occur.*

The idea that motion can come to a stop has had a considerable influence in dissociating "time" from motions, for it can be reasoned that motions can be halted while "time" cannot be stopped. The fallacy of this argument lies in the fact that the word "time" takes into account all the motions taking place in the universe, while the stopping of a specific motion is an event within a framework which itself is in motion. A car can be brought to a stop on Earth, but this is only relative to the Earth, which itself is moving relative to the Sun and other heavenly bodies. Once a motion has been brought to a stop relative to a certain framework of reference, that unit becomes a part of the framework and follows its motion. If "time" were being measured only in terms of one motion such as that of a watch, we would also be able to stop "time" without any recourse to the motion of other units; but since there are innumerable units of matter moving with respect to each other it becomes impossible to stop "time." "Time" can be stopped only if all units of matter are brought to fixed positions with respect to each other. Since this is an impossibility, we feel instinctively that "time" is a perpetual phenomenon and has been and always will be endless. But as we shall see shortly, this is not completely true.

To summarize, it may be said that the understanding of the word "time" requires the consideration of two elements, "motions" and "Man's brain." Motion is a phenomenon over which Man usually has no control, and he tries to understand it through reasoning based entirely on the principle of comparing the properties of the subject of his study with natural or arbitrary standards. For this purpose he takes an arbitrary periodic motion as his standard and determines the rate of motion of an object or the lapse of an interval between two events in the terms of this standard. But this is done only when he has an instrument available and he cares to make relatively accurate determinations. His principal way of estimating the rates of motions is by exercising a faculty inherited from his more primitive ancestors—that is, he uses the periodic cycles within his body as a standard, and he compares the rates of the incoming impulses, which represent the rate of the motion under observation, with the rates of the cycles in his body. *"Time" therefore is not an inherent property of matter or space, but a product of Man's mental activities with which he tries to understand the problems of orderly motions.*

As the study of matter is usually the study of a multitude of motions, one has to include the element of "time" in all phases of one's investigations of any natural phenomena under observation. Thus in our efforts to comprehend nature, we cannot use our usual frameworks of stationary coordinates alone. Rather, we are forced to introduce a fourth arbitrary coordinate of "time" into our calculations. The idea of fixed positions and stationary coordinates is satisfactory when one is dealing with the problems that do not include "motion." But the inclusion of the "time coordinate" becomes an absolute necessity when "matter" and "motions" are being studied jointly.

The significance of the discussion of the concept of "time" lies in the possibility of an origin and an evolution for "time." "Time" is only our invention and problem and does not play any part in the operation of the universe. But since we cannot conceive matter without "time," for us the evolution of "time" becomes an integral part of the evolution of matter. This means that there are limits to the phenomenon of "time." The beginning of "time" is very much interwoven into the conception of the origin of matter. By the same token, the end of "time" will come with the conclusion of the existence of the contemplative Man.

We shall elaborate on the above statements in the following chapters, but now we shall return to the discussions of other problems of motion.

Delineation of Motions

All motions in the universe may be divided into two main categories:

1. *The Influenced Motions:* These are the motions of units of matter that are under the influence of forces (such as gravitational and electromagnetic forces) and energies (such as electromagnetic energies) emitted by other units of matter; in almost all cases, "influenced motions" occur within "systems of equilibrium" where the courses and the directions of motions are affected by the components of the systems. Examples of this type are the motions that are observed on this planet.

2. *The Independent Motions:* These are the motions of units that move relative to each other without the motion of any one of the units being affected by the presence of other units. These "independent motions" may then be divided into two subgroups:

a. *Independent Motions by Virtue of Distance:* In this group the motions of the units are not affected by other units. They are so far away from each other that the forces and energies emitting from one unit cannot have any effect on other units. This type of motion may occur either within "systems in equilibrium by virtue of influenced motions" (e.g. the negligible effects of Sirius on the movements of a meteor in the solar system, both being components of the Milky Way) or outside these systems (e.g. the motions of two protons relative to one another in intergalactic space).

b. *Independent Motions by Virtue of the Nature of the Involved Units:* In this group the units do not emit forces and energies, and under normal conditions they are not responsive to the forces and the energies emitted by matter. The only effect that the units can have on each other is by collision. An example of this type of motion is that of photons; the motion of a photon does not affect that of other photons. The motions of the units in this group with respect to each other can be either "orderly" (e.g. in photons moving in a specific pattern) or "disorderly." In the latter case they do not possess any of the properties that are associated with organized units of matter.

The following descriptions of certain characteristics of the "independent motions" are of value in laying down the foundation for the discussion of the creation of matter in Chapter IV.

1. *If the units moving relative to each other cannot influence one another, then no periodic motion can exist between these units.* This statement applies to both subgroups of "independent motions." The essential prerequisite for the establishment of periodic motions between units of matter is the presence of effective forces such as active electromagnetic or gravitational forces. Assumably, no effective force is present among the units with "independent motions." Consequently, no periodic motions can occur among such units.

2. *When units of matter move with respect to each other in such a way as to meet the conditions described for "independent motions by virtue of distance," the units can be made to move relative to each other with velocities greater than the speed of*

light with the limit being twice the speed of light. Let us assume that two linear accelerators are constructed parallel to each other in an east-west direction twenty miles apart. Then let one unit accelerate the electrons to 90% of the speed of light from east to west and allow the second one to do the same, only in the west-east direction. The velocities of the electrons in each accelerator, irrespective of the energy input, cannot exceed the speed of light. But if we set the equipments in such a way as to prevent the motions of one set of electrons from influencing the motions of the second set, we are easily able to produce relative velocities of just below twice the speed of light. Here we have two sets of objects (electrons) whose velocities relative to Earth cannot be increased to values higher than the speed of light. However, the relative velocities between the groups can go as high as twice the speed of light; this indicates that relative velocities higher than the speed of light are possible between two objects that cannot influence each other. There are no reasons to believe that Earth is responsible for the limitations of velocities; the same experiment carried on in space would most probably give the same result and thus point to the fundamental fact that the limiting properties are universal and not simply related to Earth. **It is the space that places a limit to the rates of motions of units of matter and not the planets or stars, which are merely our reference points in the vast expanse of the "space continuum."**

3. *"Orderly independent motions," as represented by the motions of the corpuscular state of quanta (photons) in electromagnetic radiation, are closely associated with the motions of units of matter within systems whose stability is dependent on the "influenced motions" within those systems.* An example of this phenomenon is seen in the radiowaves artifically produced by the oscillation of electrons. The oscillation of electrons is regulated by the cause of orderly motions, controlled forces and energies.

4. *Since the basic laws of science have been formulated from the results of both the experiments with and the observations of the systems whose equilibrium is dependent on the "influenced motions" of their components, none of these laws are necessarily applicable to the states where "disorderly independent motions"*

predominate. No examples can be cited where "disorderly independent motions" exist. But it stands to reason that when the motions of the units do not follow any specific or predictable pattern, none of the scientific laws apply to them.

5. *If there can be no periodic motions or order in the directions or the rates of movements and no possibility of the application of the standard laws of motion to the units with "disorderly independent motions," then there would be no way of comparing the lengths of intervals between events though a succession of minor events can take place between the units.* Let us assume that we are witnessing a universe composed of quanta in their corpuscular state moving haphazardly and without any semblance of order. Assume also that the clock functions of our brains and all the rhythmic motions which are associated with the process of life are ineffective for comparison purposes. Now, visualize that in the course of their motions two quanta corpuscles get in each other's way and collide. Imagine then that another collision takes place between the newly formed pair and a third unit. Under these circumstances would there be any way in which one could determine the length of the interval between the two collisions? The answer is "no" because, within the realm of "disorderly independent motions," there are no uniform and repetitious motions with which one can compare the lapse of intervals between the two events; hence one cannot gain an idea of "time." This supports the previous statement that the phenomenon of "time" is closely related to the development of orderly motions during the creation of matter.

Motion and Space

The true measured values of lengths, time and mass are obtained only when the determinations are made either on stationary objects or the ones moving at relatively slow rates. When the measurements are made on objects moving at what is referred to as "relativistic velocities" with respect to the observer, the values obtained are quite different from the real values ascribed to the same objects in their stationary states. Time becomes dilated, lengths contract, and increases are measured in masses. These variations were originally expressed in a set of formulas collectively known as the Lorentz transformations formulas which later evolved into the theory of special relativity.

The transformation formulas revolve around one significant point: the ratio of the square of the velocity of the object to the square of the speed of light, with the velocity of the object the sole variable. The question arises: why has the velocity of light become an indispensable component of the formulas? The answer is that in all cases related to objects moving at relativistic velocities, the communication between the object and the observer is carried through the medium of light or some other form of electromagnetic wave. If light traveled at an infinite speed, the rates of the motions of the objects would not influence the measurements. But since this is not the case, the velocities of the objects become important factors in distorting the transmission of signals to the observer. That is, while the velocity of the "means of communication" between a moving object and an observer remains constant, the distance between the object and the observer is continually changing at speeds comparable to the speed of light; this causes variations in the lengths of time it takes for the signals originating at intervals (time) or positions (lengths) from the object's framework of reference to reach the observer's framework of reference. These variations cause the light (or electromagnetic waves) representing the different signals to travel varying distances and hence cause distortions of the real values.

If we follow the derivation of the transformation formulas, we find the "communication medium" accounts directly for the distortions in the measurements of time and lengths, but not of mass. In the case of mass, the knowledge of the effects of relativistic motions has been arrived at indirectly through the application of the "modified transformation formula relating to time" to the law of conservation of linear momentum. The validity of these assumptions and their resultant conclusions has been accepted largely because the predictions of the formula relating to the effects of motion on mass have been confirmed through experiments with subatomic particles. But the determination of masses in these experiments have required the measurements of momentums, inertia, and the rates of deviations in electric and magnetic fields—that is, measurements that are not dependent on the use of electromagnetic waves in the direct passage of signals. The differences between the causes of variations in those measurements of time and lengths and those of mass point to the possibility that, although time and lengths are not altered within the frameworks of the objects moving at relativistic velocities relative to "space continuum," such motions cause increases in the intrinsic structures of units of matter. *This leads to the probability that when measurements are made within the framework of reference of the moving objects, the rate of increase of mass would follow the familiar formula*

$$m = \frac{m_0}{\sqrt{1-(v/c)^2}}$$

with c the velocity of light, m_0 the rest mass and v the velocity of the object—the latter two being measured with reference to the "space continuum," using the nearby galaxies as points of reference.

Now let us proceed with the discussion of the effects of this suggested inherent increase in mass caused by motion on the problems of motion in space. The reason why we know that an object is moving is that we are able to determine its positions at intervals of time with respect to other units of matter used for reference. The same phenomenon applies in space; but it would indeed be very difficult for us to apprehend whether or not an object is in motion in space if we have no way of using the heavenly bodies as points of reference. The following situation illustrates the difficulties of determining whether or not an object in intergalactic space is in motion with respect to the "space continuum" without using the galaxies for guidance. Let us assume that an accident takes place inside a self-propelling spaceship while it is moving in the intergalactic space. The engines stop, the instruments for the observation of the galaxies are destroyed, and the occupant is knocked unconscious. Then imagine that when he gains consciousness he cannot remember whether or not he is in motion. The problems now are: has he any way of determining whether or not he is in motion, and if so, with respect to what is he moving?

In pondering the above situation one becomes aware of a number of basic questions related to the motion of objects in space:

1. Is the motion of an object in space gauged by the changes of distances between the object and units of matter as represented by the galaxies and stars or is it relative to the entire "space continuum"?

2. How does one recognize the "space continuum" without the use of stars and galaxies as points of reference?

3. If the true motion is relative to the "space continuum," is there any way of knowing that a motion is taking place without the use of stars and galaxies as points of reference?

4. Besides the relatively small quantities of matter or its products (such as radiation, gravitational forces and escaped particles from the galaxies) present in the intergalactic space, is there anything else present or is the intergalactic space absolutely and totally void?

We shall begin the discussion of these latter questions with the study of the movements of a different spaceship in the intergalactic space. We shall make three assumptions about this new spaceship and its situation: first, the solar system, instead of being at its present position in the galaxy is situated at the extreme outer regions of the Milky Way (this would make it easy for the spaceship to travel into the intergalactic space and away from the influences of the Galaxy); second, the problem of the total conversion of matter into energy has been solved and we have access to tremendous quantities of controllable energy for propulsion purposes; third, we have built a spaceship capable of using the energy from the annihilation of matter at any rate we desire. The spaceship is then used to investigate the problems of motion in intergalactic space far away from the Galaxy, where the latter's forces and energies will not interfere with the experiments.

The spaceship is launched toward a distant galaxy "A" and is allowed to move at relativistic velocities away from the Milky Way until it is so far away that for all practical purposes it becomes independent of the Galaxy's gravitational forces. The rate of recession from the Milky Way and the approach toward the galaxy "A" are measured regularly and recorded.

The following predictions may be made about the motions and the properties of the spaceship through the application of the principles of the laws of motion and the stipulations of the special theory of relativity:

1. No matter what amount of energy is expended, the spaceship will not be able to move at a rate equal to or faster than the speed of light with respect to the two galaxies as well as the "space continuum."

2. As the velocity increases the masses of the ship and its contents increase according to the proposed modifications.

3. The increase in mass in response to velocity may be determined within the ship through the use of instruments which can estimate the rate of increase of a standard mass with a known inertia by measuring its inertia; in this way the rate of the motion of the spaceship (while moving at relativistic velocities) relative to the "space continuum" may be measured without any recourse to using the Milky Way or the galaxy "A" as the points of reference.

4. While moving in a straight line toward the galaxy "A," the engines can be adjusted to make the spaceship move at a constant velocity. Extra amounts of energy are needed to change the direction of the movement of the spaceship; the required amount of energy is dependent on the mass of the spaceship and on the degree of the desired deviation from the straight line.

From the situation of the spaceship just mentioned, the following conclusions may be drawn:

1. *The intergalactic space cannot be considered as totally and absolutely void as long as the motions of the objects moving in the intergalactic space produce results confirming the predictions of the laws of motion and the stipulations of the special theory of relativity.* In a state of absolute void, a moving object is entirely independent of its surroundings and as a result is free to move at any imaginable velocity. The fact that the "space" puts up a resistance to the propulsion energies of the spaceship engines (phenomenon of inertia) gives strong indications that the intergalactic space contains some universal factor that, though imperceptible to us, has the property of responding to the motion of objects relative to the "space continuum."

2. *The imperceptible factor in space which shows its presence when objects move at relativistic velocities in intergalactic space is also present in interstellar space and consequently within all objects.* We know that in spite of the amount of energy used, subatomic particles cannot be made to move at velocities equal to or greater than the speed of light in accelerators. Thus the same factors that prevent the speed of objects from reaching values higher than the speed of light in the intergalactic space repeat the same function within matter and interstellar space. Therefore, we should expect that these factors (which also bring about increases in the mass of objects moving at relativistic velocities in space) would be present in the interstellar space and the space between the subatomic particles within matter. Henceforth these factors will be referred to as "space-prim-atoms," and for the present let us accept each primaton as being the same as a quantum in its corpuscular state but possessing no energy.

3. *The "space-primatons" move at the speed of light with respect to the "space continuum."* There is an established limit for the speed of any object moving relative to the "space con-tinuum" and apparently the control of this speed comes from space itself. Thus it seems reasonable to assume that the "space-primatons" move at the same maximum velocity and because of that they prevent any object from moving faster than the speed of light relative to the "space continuum."

4. *The "space-primatons" always move along straight lines.* This is part of the assumption that the "space-primatons" are the corpuscular state of quanta without energy. Accordingly, the resistance offered by space to the change of direction from a straight line of a free-moving object can be attributed to the movement of "space-primatons" along straight lines.

5. *The demonstration of the property of inertia in an object requires the disturbance of the established equilibrium existing between the object and the "space-primatons."* An object stays in its state of rest until it is made to move by the action of an extraneous force or energy. The resulting motion will continue along a straight line until new forces and energies cause a change in the direction or the rate of the motion. Two questions now present themselves: why does an object require the

application of force or energy to change its stationary state within an arbitrary framework of reference and why is the resulting motion always along a straight line? The hypothesis presented here explains this phenomenon in terms of resistance to change by the state of the equilibrium established between the "space-primatons" and the object. As long as the object is stationary within a framework of reference (which is usually moving at a steady non-relativistic rate relative to the "space continuum") or moving at a constant non-relativistic speed along a straight path, the equilibrium consists of the "space-primatons" constantly moving in and out of the object. This brings about a state of balance between the object and the "space continuum." The only time this equilibrium is disturbed is when forces and energies from outside interfere with these established uniform motions and bring about changes in the state of equilibrium of the object with respect to the "space continuum."

6. *"Space-primatons" differ from the corpuscular state of light quanta in two significant respects: they do not possess energy, and although their motions are along straight paths, they move in any possible direction relative to the "space continuum."* We have already discussed the absence of energy among the "space-primatons." The cause of this condition will be discussed in the next chapter. In regard to the unpredictability of the directions of their motions, the "space-primatons" are unlike the photons which usually originate from a source and spread out in an orderly manner. Instead, the "space-primatons" do not have a source, and thus they move in any possible direction without any specific pattern.

7. *The increase in the mass of objects moving at relativistic velocities indicates that the "space-primatons" can be converted to matter.* The theory of the creation of matter in Chapter IV is based on the interchangeability of certain types of primatons with matter, but at this juncture it should be pointed out that the creation of matter could not have begun with the conversion of "space-primatons" into matter. The reason for this statement is obvious: "space-primatons" do not possess energy, and therefore, the creation of matter could not have come from such a neutral source. It should be remembered that the conversion of

"space-primatons" to mass takes place only under specific and unusual conditions which could not have existed during the creation of matter.

8. *If the "space-primatons" do not convert to mass under normal conditions, are devoid of energy, are distributed throughout the "space continuum," and are not the source of matter, then they must be a product of the evolution of matter.* This statement indicates that there were no "space-primatons" prior to the creation of matter and they came into being after the appearance of matter.

9. *If the "space-primatons" are perpetually moving at the speed of light in all possible directions, then the "space continuum" may be visualized as a matrix composed of "space-primatons" darting about in any possible direction with the galaxies acting as its recognizable points of reference.* The fundamental point in this statement is that the total sum of all the motions of "space-primatons" provides us with a medium without any geometric configuration with which the rate of motion of any individual object may be compared. Accordingly, the true motion of any object in the universe is its motion relative to the "space continuum" and not to any artificially constructed framework of reference. According to this interpretation of space, the frameworks of reference are only our guides for orienting ourselves with respect to the "space continuum" and consequently cannot be regarded as the causes of the observed properties of matter.

It should be pointed out that the concept of the "space continuum" is not the same as the "ether hypothesis" which was prevalent previous to the Michelson-Morley experiment. In that hypothesis ether was supposed to be the medium through which light moved. The "space continuum" is visualized as a matrix composed principally of corpuscles possessing neither energy nor force and moving with the speed of light. The motion of light through space is not dependent on the presence of the "space-primatons" in space.

Force and Energy

In classical terms, "force" is defined as the cause of the change in the state of rest or motion of a body, and "energy" as the capacity for performing work. However, on close scrutiny one finds that at times these properties overlap. For example, in the cocking of a spring rifle, the potential energy is stored in the spring, yet after the pulling of the trigger it brings about the acceleration of the bullet which, according to definition, is the function of force. There are a number of examples of this type in which either word, "force" or "energy," may be applicable. The reason for the lack of precision lies in the different ways these terms have been used through the years.

To facilitate the coming discussions, we shall qualify the concepts of "force" and "energy" in the following way: all "forces" and "energies" shall be subclassified into two general groups, 'intrinsic" and "mechanical." These shall have the following properties:

1. Intrinsic forces and energies do not need contact as a means of transfer from one source to another; mechanical forces and energies cannot be transferred without the direct contact between units of matter.

2. Intrinsic forces and energies obey laws related inversely to distances, a phenomenon not applicable to mechanical forces and energies.

3. When mechanical forces and energies are applied to objects, the resulting motions are always in the same direction as the direction of application. The intrinsic forces always cause motions in the opposite direction to the direction of application, and the intrinsic energies cause motions in the same direction as the direction of application.

Nuclear, electromagnetic and gravitational forces are intrinsic forces; electromagnetic energies are the only known intrinsic energies. But, as we shall see later, the possibility exists that under certain conditions non-wave radiations are also produced. The mechanical forces and energies include all the forces and energies conveyed and transferred by objects through contact. It should be pointed out, however, that the energies and forces transmitted by mechanical means always originate either from intrinsic forces or intrinsic energies.

One reason why the definitions of "force" and "energy" are biased toward mechanics is that at the time these words were being defined only mechanical methods were being employed for the measurements of intrinsic forces and energies. Even today the principal means of perceiving the presence of intrinsic forces and energies and quantitatively determining their strengths is through the use of mechanical devices. Indeed it may be said that Man has no faculty for the direct and immediate recognition of any of the intrinsic forces and most of the energies in the electromagnetic spectrum.

The following are the three principal differences between the properties of intrinsic energies and intrinsic forces:

1. The direction of motion caused by energies on objects is away from the source of energies; the direction of motion caused by forces is toward the source.

2. All of the known intrinsic energies are generated in series of waves with such properties as reflection, refraction, diffraction, etc. Intrinsic forces do not possess these properties.

3. The generation of intrinsic energies requires an input of energy at least equal to the output; intrinsic forces continually emanate from their sources without the need for the input of any force or energy and without the loss of mass. Emanated forces are always present around the source and when they get the opportunity to act on a freely moving, responding object, they cause it to accelerate. Thus the ever-present forces always have the potential to become converted to kinetic energy, and this act can be repeated infintely without any material effect on the source of the force.

Under normal conditions, intrinsic forces produce almost exactly the same effects on objects as intrinsic energies; the main difference between the two is the direction in which the object is caused to move. For example, if we permit a rocket to ascend very slowly above the atmosphere and then adjust the engines to make it hover stationary over the Earth, we have created a condition in which the intrinsic energies from the engines and the gravitational forces from the Earth have come into equilibrium and are neutralizing the effects of each other. That the forces are performing exactly the same functions as the energies suggests that the basic mechanism for the action of the intrinsic forces and energies on the atoms of the body of the rocket is the same.

There is not yet any experimental data regarding the velocity of the intrinsic forces, but the indications are that, similar to electromagnetic waves, they move with the speed of light. This deduction is based on the similarities that exist between the two. They both obey the inverse square law of distance. Like electromagnetic waves, intrinsic forces can be converted into kinetic energies under appropriate conditions; also they can be directly neutralized by intrinsic energies.

If it is proved that the intrinsic forces travel with the speed of light, then a very fundamental relationship develops between "space-primatons," intrinsic energies, and intrinsic forces. We know the energy responsible for the photoelectric effect is carried by the corpuscular state of quanta in the form of photons; it has been proposed that "space-primatons" are the "no energy" quanta in their corpuscular state. Using the principle of symmetry in nature, it may be reasoned that the intrinsic forces are also composed of quanta in their corpuscular state in the "symmetrical image" of the photons. On the basis of these arguments it may then be concluded that the corpuscular state of quanta is the base for the three, with the exhibited difference in their properties being caused by the source from which they are generated. Since it is not practical to use the phrase "quanta in their corpuscular state" continuously and the expression "photon" applies only to electromagnetic radiation, from now on we shall use the term "primaton" for this fundamental unit in the universe.

We have now reached the critical stage of having to answer: what is meant exactly by the "corpuscular state of quanta"? The photoelectric effect has established the particle nature of light quanta and the Compton effect has confirmed these findings and interpretations. In spite of the importance of this phenomenon, no attempts have been made to give a clear description of the true nature of a "quantum in its corpuscular state." The reason for this avoidance, of course, has been the dual nature of light, for under certain conditions light acts as particles and under other as waves. Nonetheless, it we are talking about a "particle state" we are referring to an existing entity with measurable dimensions. We shall discuss the relationship of the wave and corpuscular nature of light in the next chapter. At this stage of the discussion we are interested only in the corpuscular nature of quanta which we have called "primaton." *We hereby define a "primaton" as the smallest known spherical-shaped particle with a definite but as yet undetermined diameter.*

A primaton can exist in one of three states. These differ from each other on the basis of one single property: the direction in which the primaton causes a free object to move after it imparts its effects. An "energy-primaton" makes the object move in the same direction as the linear motion of the primaton, a "space-primaton" does not have any effect, and a "force-primaton" causes the object to move in the direction opposite to the direction of the linear motion of the primaton. As energy is recognized not only in terms of values but in terms of direction of effects, it may be said that, in a standard coordinate system, the energy-primatons possess positive energy, the space-primatons have no energy, and the force-primatons exhibit negative energy.

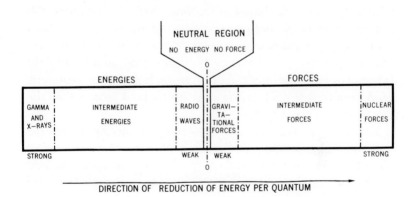

Figure 1. THE FORCE-ENERGY SPECTRUM

On the basis of the amount of energy carried by a primaton, a general spectrum may be constructed (Figure 1) covering the entire range of the intrinsic forces and energies in the universe. The diagram is self-explanatory; however, attention should be drawn to the following points:

1. The spectrum is symmetrical in its pattern.
2. The spectrum terminates in an abrupt fashion at both ends. This indicates the existence of maximum limits for the strengths of positive and negative energies carried by primatons.
3. There is no sharp or distinct boundary between the weak forces and energies; both gradually fade into the no-energy state toward the center. On the energy side, this slow transition is supported by experimental data, but due to a lack of detecting instruments, no information is available as to whether or not this assumption is also valid with respect to the gravitational forces.
4. Whereas in the electromagnetic radiation section no vacant area exists in the spectrum, in the force section there are apparent vacancies between the known forces. So far we know of three major types of forces: nuclear, electromagnetic and gravitational. Each has its own specific strength and each occupies a certain zone in the spectrum. It should be noted that the broken lines on the diagram are not meant to indicate sharp separation of the force zones, but are for the purpose of showing the approximate positions of the known forces in the spectrum.

Before we proceed any further, an understanding should be reached in the use of certain words. In this book we shall use the following terms and symbols:

matter—any existing and recognizable material in the universe irrespective of whether it is made of antimatter or the familiar matter on Earth.

+*matter*—the type of matter whose principal components are protons, neutrons and electrons.

antimatter—the type of matter composed principally of antiprotons, antineutrons and positrons.

+*H*—hydrogen composed of protons and electrons.

antiH—antihydrogen.

force—intrinsic force.

energy—intrinsic energy.

Note 1. Most of our discussion will be regarding forces and energies emanated from matter. Hence the words force and energy will refer only to the intrinsic type. Should we need to refer to specific forces or energies we shall use the descriptive term, e.g. potential energy or gravitational force.

Note 2. We shall use the word antagonistic to denote the two states of matter in relation to each other. For example, +*matter* gases are antagonistic to *antimatter* gases.

Note 3. We shall designate +*matter-antimatter reactions* for the reactions taking place between +*matter* and *antimatter*; the symbols for reactions between hydrogen and antihydrogen will be +*H-antiH reactions.*

III

FORCES, ENERGIES, SPACE
AND REDSHIFT

All fields around every body or particle should be considered as something real. —*Albert Einstein*

In the previous chapter we drew attention to the possibility of forces being the symmetrical counterparts of energies in a force-energy spectrum. The validity of this conclusion is of course dependent on experiments which can show whether or not forces travel with the speed of light. Pending the completion of these experiments, we will assume the proposition as valid and accept the concept that under certain conditions the primatons become the conveyors of forces. If this be the case, then the problem will be to determine the specific conditions under which the primatons become converted to force or energy, or transform to the totally neutral state of space-primatons. It is proposed that the relationship between the rotary and the linear motions of primatons determines whether these corpuscles will be carriers of energy, force or zero force-energy.

Modes of Action of Force- and Energy-Primatons

A principal concept in this book has been that all of the properties of matter can and should be explained in terms of motion; forces and energies are no exception to this rule. A primaton, as depicted in these propositions, is an existing entity with a specific volume and thus it has the possibility of developing two types of motion: linear and rotational. It is suggested that when a primaton comes into contact with an object, its linear motion

31

imparts positive energy to the object, while its rotational motion affects the object by imparting a negative energy to it. As only one value is possible for the linear motion—the speed of light—the total amount of energy of a primaton can be deduced by combining the amount of the negative energy caused by its rate of rotation with the constant amount of energy produced by its linear motion. The latter is approximately equal to the energy of one gamma-ray primaton.

Figure 2 shows the manner in which a force-primaton causes a subatomic particle (a proton here) to move in the direction opposite to that of the linear motion of the primaton. Theoretically a primaton should be able to rotate in any direction with respect to its linear motion. But, as will be shown shortly, the subatomic particles from which the force-primatons and energy-primatons emanate always cause them to rotate along the axes perpendicular to their linear motions. In the diagram the axis of rotation for the primaton has been drawn to be perpendicular to both the plane of the paper and the direction of the linear motion of the primaton. The two types of motions produce two kinds of effects: (1) the linear one imparts to the proton a motion in the same direction as that of the primaton; and (2) the rotational motion causes the proton to move in a direction opposite to that of the linear motion. In this case there is greater negative energy than positive energy and hence the primaton is causing the proton to move in the opposite direction to that of the linear motion of the primaton.

Since the rotational energy of the primaton pictured in Figure 2 is greater than its linear energy, it has caused the proton to move in a direction opposite to its own linear motion. Thus it is a force-primaton. Had the primaton caused the proton to move in the same direction as it was proceeding, it could be deduced that the positive energy resulting from the linear motion was greater than its negative energy caused by the rotational motion. This would have been an energy-primaton. Had the proton not moved when the primaton came in contact with it, the primaton would have shown itself to be a space-primaton—that is, a primaton with no energy due to the fact that the distances covered by its linear and rotational motions are equal.

According to the above considerations, the amount of energy

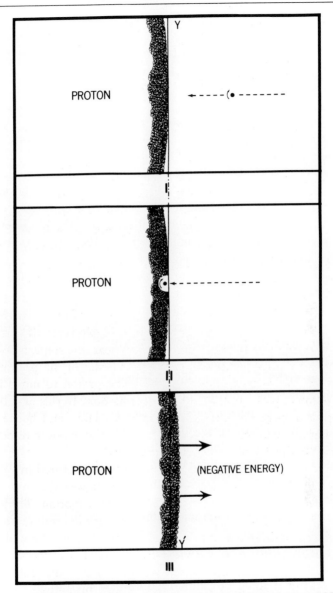

Figure 2. DIAGRAM SHOWING THE WAY THE ROTATION OF A FORCE-PRIMATON CAUSES A FREELY MOVING PROTON TO MOVE IN THE OPPOSITE DIRECTION (CROSSES THE REFERENCE LINE YY') TO THAT OF THE LINEAR MOTION OF THE PRIMATON. In this case the rate of rotation is high enough to neutralize the positive energy and also impart a negative energy to the proton. Frame I: The approach of the primaton. Frame II: The collision of the primaton with the proton. Frame III: The direction of motion imparted to the proton.

that any primaton can transfer to an object is equal to the amount of energy caused by its linear motion plus the negative energy caused by the rate of rotation of the primaton. In mathematical terms,

$$E = k(c\text{-}n\pi d)$$

with E being the energy carried by a single primaton, k a constant, c the velocity of light, n the number of rotations per second, and d the diameter of the primaton (a constant of unknown value). The formula provides the following information:

1. If $n\pi d$ is equal to zero, E will be at its maximum possible value. Most probably gamma-rays belong to this group.

2. As long as $n\pi d$ has a value smaller than c, the primaton is carrying positive energy, but the greater the rate of rotation the smaller the amount of positive energy.

3. If $n\pi d$ is equal to c, the primaton is without energy (space-primaton).

4. If $n\pi d$ has a higher value than c, the primaton is transporting negative energy, and the greater the rate of rotation the stronger the force.

When one attempts to calculate the rate of a primaton's rotation, one arrives at fantastically high figures. For example, even without knowing the value of a primaton's diameter, it can be deduced that a space-primaton rotates at such a speed that a point on its equator covers the distance of 299,790 kilometers per second. This is an almost unbelieveable value, but since there are no factors in space which can interfere with this rate of rotation and none of the laws of motion are applicable to the motions of primatons, no reason exists for the rate of rotation of the primatons not to reach such values or even higher.

Primaton Subatomic Particle Relationship

Energy and force are both produced by matter, and since matter is composed of multitudes of subatomic particles, the transformation of the primatons from one state to the other becomes a consequence of their encounter with the subatomic particles. But we also know that none of the subatomic particles are solid wholes. It has been shown repeatedly that all of the known particles (with the exception of the elusive neutrino) can be broken down to smaller units, built up to larger than normal particles, or converted directly to energy. Yet, no matter which subatomic particles are used in any of the experiments, all are convertible to energy, generally high-energy rays. According to the presented hypothesis, energy is one of the states of primatons. Therefore, each subatomic particle may be regarded as a conglomerate of a large but specific number of primatons in a neutral state but with the potential to be transformed to energy-primatons.

This is the summary of the proposed concepts. In singular form the primatons can assume any one of three states of force, energy or zero-force-energy; in the conglomerate form, the primatons are in the subatomic particle state. In order to be stable, the subatomic particles can have only one of the two masses, that of the proton (or antiproton) or that of the electron (or positron). The restriction of mass of stable particles means a definite number of primatons in each of the particles. We do not know why stability and a specific mass go together, but the velocity of light, the diameter of the primatons and the rate of the spin of the particles probably have had an important role in this selection. Finally, the cause of the emanation of forces and energies from matter may be attributed to the collisions of primatons with subatomic particles.

It is beyond the scope of this book to discuss the nature and the structure of subatomic particles, and how and why they hold together to form different atoms. For our purpose here we need only a simplified picture of the stable subatomic particles. Accordingly, a proton (or antiproton) is understood to be a central spinning core surrounded by a cloud of mesons, and an electron (or positron) a very small, solid spinning core. We shall then consider each core (including those of mesons) as a conglomerate of tightly packed primatons with the potential to be converted to energy. Our main concern will be to indicate in a general way how a field of force is produced around a particle and describe the mechanism for the generation of lights of specific wavelengths by oscillating electrons.

The process of the transfer of force or energy from a primaton to a subatomic particle may be regarded as the imparting of the positive or negative energy to the particle when the two come in contact. When *force-* or *energy*-primatons collide with a sub-atomic particle, they penetrate the particle; this is due to the fact that they possess either negative or positive energy. Space-primatons, being neutral, cannot pass through the outer layers; they are bounced back by the effects of the spin and the elastic properties of the particle, and leave no effects on the particle. The penetration of an *energy-* or *force*-primaton into a free-moving subatomic particle causes the particle to accelerate; in the process the primaton loses its positive or negative energy and becomes a neutral component of the particle. But the particle is in a state of equilibrium with its surroundings and has to preserve that state; thus it disposes one of its own primatons into space in the form of the no-energy state of a space-primaton.

Generation of Forces

The released space-primatons move away from the subatomic particles with the speed of light. If the particle is traveling in space, these released neutral corpuscles join the countless numbers of space-primatons in space and become part of the space continuum. But if the particle is part of a mass, in most cases the released space-primatons collide with other subatomic particles along their paths. It has been stipulated that these neutral primatons cannot penetrate the encountered particles and are bounced back. It is now proposed that this process transforms these primatons to force-primatons. We shall focus on their transformation to gravitational and electromagnetic forces only; the "nuclear" and the "weak" interactions are related to the little-understood mechanisms inside the nucleus and are probably beyond the effective participation of space-primatons. But before going any further let us take a look at some of the pertinent properties of forces.

The most important fact about a field of force is that the forces do not seem to obey the laws of conservation in the manner of the electromagnetic energies. In order to emit electromagnetic radiation, a continuous input of at least an equivalent amount of energy is needed. In contrast, the forces are produced around subatomic particles without an input of energy or force and remain unimpaired no matter how many times the emanated forces are converted to kinetic energy. The application of either electromagnetic energies or forces to a particle in space causes the particle to accelerate; the direction of the resulting motion caused by energy is different from that of force, but in both instances the particle gains kinetic energy. In the case of electromagnetic energies, the transferred energy has its origin in the energy input. But forces are generated continuously without a need for a balancing input of energy or force and without any material effect on the source. Apparently, this is in contradiction to the stipulations of the laws of conservation. We shall discuss the relationship of the production of forces and the laws of conservation very shortly, but from what has already been described we can draw the conclusion: *the cause of the production of a field of force around a particle is a mechanism within the particle which has the capability of converting the neutral state of space to a field of force without infringing on the laws of conservation.*

The above conclusion implies that as a result of the collisions of neutral primatons with subatomic particles there is a continual conversion of space-primatons to force-primatons. However, the mechanism of this transformation is rather obscure, the lack of clarity coming from our insufficient knowledge of the exact structures of the elementary particles and the manner of the maintenance of their equilibria within matter. This means that our discussions on this subject should be in general terms. Accordingly, it is suggested that the increase in the rates of rotations of space-primatons is caused by the spins of the particles with which they collide. Now, if we assume the collisions are taking place on spherical surfaces, then the polar regions of the particles produce the weaker forces and the stronger forces are generated from the main surface; the closer the point of contact is to the equator, the stronger the produced force-primaton. It is proposed that the forces generated at polar regions are the gravitational forces and the ones from the main surface are the electromagnetic forces. This proposal is supported by the measurements and calculations regarding the ratio of the total strengths of electromagnetic to gravitational forces generated by subatomic particles. It has been shown that the electromagnetic forces between a proton and an electron is $2.3x10^{39}$ times greater than the gravitational forces between the two, indicating the production of far greater numbers of stronger electromagnetic force-primatons around a subatomic particle than the weaker gravitational forces.

One may draw the following conclusions from these interpretations:

1. The collision of a space-primaton with a subatomic particle results in its transformation to a force-primaton. The converted primaton leaves the spinning particle with a newly acquired rotation around an axis which, similar to all primatons, is perpendicular to the path of its linear motion.

2. The electromagnetic forces being emitted from a subatomic particle are a mixture of force-primatons of different strengths, ranging from the comparatively strong ones produced at the equator to the relatively weak ones produced at the border between the electromagnetic and the polar gravitational regions. The measured strength of the electromagnetic forces is the average of the strengths of all the force-primatons generated from the electromagnetic force region on the particle.

3. The strengths of the gravitational force-primatons vary within a narrow range. They are not all exactly of the same strength, for the strength depends on the point of collision on the area around the pole.

4. There is no sharp boundary between the strengths of the strongest gravitational and the weakest electromagnetic force-primatons.

5. The separation of the electromagnetic force-primatons from the gravitational primatons originating from a proton in an atom of hydrogen is caused by the neutralization of the former by the electromagnetic forces originating from the atom's electron. Under natural conditions the two forces with opposite charges are continually neutralzing each other. The only portion of the field around the proton that cannot be neutralized by the oppositely charged forces from the electron are the ones generating in the polar regions; the primatons generated in the polar regions escape into space as gravitational forces.

6. The gravitational force-primatons, though weak in strength

and relatively small in number, are cumulative. They represent all of the force-primatons a neutral mass can emit to its surroundings.

7. The strength and the effective range of a field of force is governed by two principal factors: the strength of the negative energy carried per force-primaton and the number of force-primatons that have escaped the equilibrium of the atoms. The two complement each other. For example, if there are relatively small numbers of strong force-primatons in the field, it is a strong field but short ranged; on the other hand, if there are a large number of weak force-primatons in the field, it is a weak field but has a long range.

Two questions present themselves with regard to the formation of force-primatons from space-primatons: (1) will this type of conversion have any effect on the subatomic particles?; and (2) can one reconcile such conversions with the laws of conservation? The answer to the first question is: since the space-primatons possess neither energy nor force, collide with the particles from all directions and do not penetrate them, the structures and motions of the particles cannot be influenced by the collisions. The second question can be answered in terms of the pattern of propagation of the produced force-primatons. The space-primatons arrive from all directions, and after their conversion they propagate back uniformly in all directions. Statistically speaking, for every force-primaton produced at a point on the surface of a particle, an equivalent one is produced at its opposite side. But after their impacts these two primatons move in opposite directions; alone, each has the potential to impart kinetic energy to a particle, but the combined effects of the two would be zero energy potential. We may then conclude: *the sum of all the negative energies produced by a particle within any specified period is zero.* Consequently, in the formation of force-primatons from space-primatons, the law of conservation of energy is preserved.

Generation of Electromagnetic Energies and Waves

Our main interest in the following discussion will be the explanation of the nature of light and the mechanism for its generation from a simple atom exemplified by an atom of hydrogen.

Experimental results have shown electromagnetic waves to be of dual nature—that is, under certain conditions they have wave properties and under other conditions the characteristics of particles. The waves are considered to be of a transverse type, and their propagation is compared to that of the waves produced in one plane along the length of a rope attached to a hook. The reason for the choice of this pattern is the phenomenon of polarization. However, the difficulties in these theories are twofold. In the first place, the wave pattern is explained in terms of the pattern of motion of objects made of matter; consequently the true nature of the wave remains obscure. In the second place, it is assumed that light can be either in the corpuscular state or the wave state but not both at the same time, the assumption being based on the fact that the two properties cannot be demonstrated by a single experiment.

The hypothesis in this book considers light (or any other electromagnetic wave) as a series of concentric spherical shells expanding from a source with the velocity of light. Each spherical shell is then regarded as being composed of energy-primatons. Under this concept the corpuscular and the wave natures of light occur simultaneously, the former being the real nature of radiation and the latter the mode of its propagation. The polarization phenomenon is then explained on the basis of the direction of rotation of primatons along their paths.

In order to explain this hypothesis we shall make the following stipulations based on established principles:

1. The electrons in the atoms of hydrogen oscillate during their transition from one energy level to a lower level. The oscillation causes the disposal of the excess energy stored in the electron (equal to the difference between the two energy levels) in the form of energy-primatons.

2. The electrons of the atoms in a source of light that are going through identical transitions in energy levels "oscillate in phase." By "oscillation in phase" we mean these electrons oscillate exactly at the same rate and reach their peaks and troughs together precisely at the same time. This stipulation is based on the principles of the resonance phenomenon observed in tuning forks of the same natural frequency, and the resonance absorp-

tion phenomenon in lights of the same frequency. The support for this premise comes from the fact that we can produce interference patterns; if there were no "oscillation in phase," the waves from a number of electrons in a source would overlap and prevent the formation of such patterns.

Oscillation is a special type of motion produced in an object with limited freedom of motion after it receives energy from outside. The function of oscillation is to dispose of the energy so that the system can return back to its state of equilibrium at the median point between the two opposite maximum oscillatory positions. In an atom of hydrogen, the oscillation responsible for the emission of light occurs during the transition of the electron from a temporary high energy level to a lower level. In this transition the electron is disposing excess energy-primatons acquired from external sources; these energy-primatons have brought about the differences in the energy levels and have caused the oscillation. Here the disposal of energy occurs by the ejection of the energy-primatons at that point in the oscillation where the direction of motion is changed, the amount of energy per ejected primaton being determined precisely by the rate of oscillation. But the rate of oscillation as well as the velocity of motion of the electron in the course of its oscillation are functions of the difference between the two energy levels. Accordingly, the energies of the energy-primatons and the period between the two successive emissions are both functions of the same factor—the difference between the two energy levels.

It has been stipulated that in a source of light all the electrons passing through identical energy-level transitions oscillate in phase. This means that a set of oscillations in phase by a number of electrons produces a large number of energy-primatons in the form of an expanding spherical shell, with each primaton carrying a specific amount of energy; but this also means that sets of such oscillations follow each other continually at exact and regular intervals, resulting in the emission of concentric shells which move away at precisely equal distances from one another. Both the energy of the energy-primatons in the shells and the distances between the shells (wavelengths) are functions of the rate of oscillation of the electrons—which in turn is a function of the difference between the two energy levels. Consequently, a monochromatic light possesses two distinguishable properties, that of wave, which is merely the pattern of intermittent emission of energy primatons, and that of particles, each corpuscle carrying a definite amount of energy. If we measure distances between the shells by interferometric techniques, we are studying the wave nature of light; and if we determine the amount of energy, as is done by photoelectric methods, we are working with the particle nature of light. The two properties are together all the time, with the most significant part of the phenomenon being that *under natural conditions, in electromagnetic waves, a specific wavelength is always associated with a specific amount of energy per energy-primaton.*

We shall now attempt to explain the third phenomenon associated with the nature of electromagnetic waves, polarization. Two orientations can be assigned to the axes of rotation of energy-primatons. The first one is always perpendicular to the path of the linear motion of primatons. The second is with respect to an arbitrary horizontal line perpendicular to the path of the linear motion of any of the primatons and passing through its center. The frame A of the diagram in Figure 3 represents a number of energy-primatons in a non-polarized light emitted from a source behind the frame. Here the axes of rotation of the primatons, while remaining perpendicular to the path of light, can assume any position with respect to the horizontal line. The process of polarization then may be described as the removal of those energy-primatons with axes of rotation at certain orientation with respect to the horizontal line by the atoms of the polarizing medium. In the B frame of Figure 3, the primatons with horizontal axes have been filtered out. Now, if we place a second polarizing medium in the path of the light in frame B with its polarizing axis perpendicular to that of the first one, the remainder of the primatons will also be removed and no light will get through the frame.

Phenomenon of Redshift

The redshift in the spectral lines has been observed under three separate and apparently unrelated conditions: (1) the redshift related to a light coming from a receding source (Doppler effect);

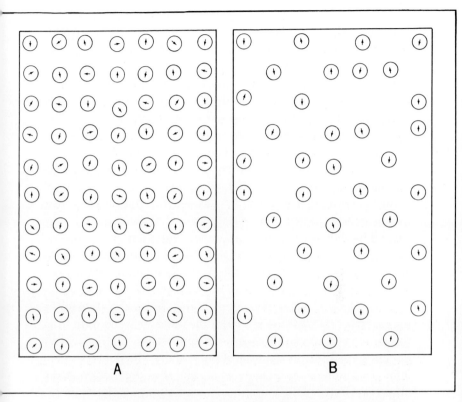

Figure 3. DIAGRAM SHOWING THE MANNER OF REMOVAL OF ENERGY-
PRIMATONS WITH CERTAIN AXIAL ORIENTATIONS BY A PO-
LARIZING MEDIUM. Frame A: Non-polarized light with random
orientations of the axes of rotation of the energy-primatons. Frame B:
The same light after having passed through a polarizing medium. Note
that only the energy-primatons with certain axial orientations have
passsed through.

(2) the redshift which is a function of distance between the source and the observer; and (3) the redshift caused by the effects of gravitational forces. We shall attempt to explain these observations in terms of the presented hypotheses and then try to find if any correlation exists between the three types.

The first point that should be considered is the methodology of determining the redshift. The determinations have principally been carried out by the use of spectrographic methods— techniques which do not make direct measurements of wavelengths. The true values of wavelengths can be measured only by interferometric methods. The spectrographic techniques cause the separation of lights of different wavelengths, not because of being sensitive to variations in wavelengths, but because the instruments respond to the variations in the amounts of energy carried by the energy-primatons. Since under normal conditions each specific wavelength is associated with a specific amount of energy per energy-primaton, the spectrographic determinations indirectly give accurate values of the wavelengths.

In the Doppler effect, the motion of the observer relative to the source causes the shift toward red if the observer is receding from the source and toward blue if he is approaching it. Let us first discuss the redshift. It has been stipulated that the dual nature of light is a simultaneous phenomenon, and thus the motion of the observer relative to the source has two simultaneous effects on the intercepted light: it results in an increase in the wavelength as well as a proportionate reduction in the energies of the energy-primatons. In the first case, during the interval between the emission of two sets of energy-primatons, the observer moves a certain distance (function of the velocity of recession) away from the source; hence the distance between the two concentric shells (wavelength) registered on the observer's instrument increases by the factor $\frac{v + c}{c}$ given in the Doppler effect formula. In the second case, the receding observer receives the energy-primatons at velocities equal to $\frac{c}{c + v}$—that is, with proportionately reduced energies. In other words, the same phenomenon which causes the increase in wavelengths also causes a compensating reduction in the energies of the energy-primatons; as a result, the relationship between a wavelength and the energy associated with that wavelength is maintained. We can use the same reasoning to show that in a source approaching an observer the reduction of wavelength and the increase in the energy of the energy-primatons also go hand in hand. To summarize, *in the Doppler effect, the increase or the decrease in the wavelength corresponds respectively with the decrease or the increase of the energies of the energy-primatons, with both properties being affected precisely by a factor which is a function of the rate of movement of the source relative to the observer.*

The redshift, which is a product of distance, is caused by a different mechanism than the redshift due to Doppler effect. At the risk of being repetitious, we note again that at the time of emission of light a predictable relationship exists between the value of energy per energy-primaton and the wavelength. However, after leaving the source, light has to travel through the space continuum containing an innumerable number of space-primatons darting in all directions. Every so often the energy-primatons collide with the space-primatons, and depending on the angle of contact, the space-primatons transfer a portion of their higher rate of rotation to the energy-primatons; in the process some of the energy of the energy-primatons is lost to the space-primatons. If the light has traveled a relatively short distance before reaching our instruments, the loss of energy is infinitesimal and cannot be detected, but after a long journey through space the loss of energy becomes measurable. To summarize:

1. Since the loss of energy per energy-primaton in the light traveling through space is a function of the number of collisions between the *space-* and *energy-*primatons (which in turn is a function of distance), the amount of redshift is proportional to the distance between the source and the observer.

2. In a redshift caused by distance, the wavelength (distance between the concentric shells) is not affected by the loss of energy. *Consequently, the exact normal relationship between the amount of energy per energy-primaton and the wavelength is not maintained.*

3. Since the amount of loss of energy per energy-primaton is a question of chance encounters, there will not be exactly the same amount of reduction of energy for all the primatons. Accordingly, the spectrographic lines that have gone through

the "distant redshift" will not be as sharp as when the same amount of redshift is caused by the Doppler effect, and the farther the distance the less the sharpness of the lines.

The third type of redshift, as predicted by the general theory of relativity, is caused by gravitational forces. According to the explanations given in this book, the phenomenon is fundamentally the same as the redshift caused by distance; in this case, however, instead of the space-primatons being the causative agents, the gravitational force-primatons are responsible for the shift. But force-primatons have higher rates of rotation than space-primatons; consequently, each encounter reduces the energy of the energy-primatons by a greater factor. Also, since gravitational force-primatons are at their maximum concentration in the vicinity of masses, the opportunity for encounters are much greater in the regions around the great masses than in space.

Suggested Experiments for Testing the Validity of the Proposed Interpretations

If force-primatons are the symmetrical counterparts of energy-primatons, then a monochromatic light passing through an intense radiation zone should have a blueshift in proportion to the intensity of radiation. We believe this to be valid and suggest it to be used as one of the test cases for the proposed hypothesis. However, in order to satisfy the experimental conditions, the radiation should come from one direction and at its maximum effective angle, probably perpendicular to the path of light. In such a case we should not only get a blueshift but light should also bend away from the source of radiation in proportion to the shift. The best way to check this conclusion seems to be by experiments conducted in laboratories. The measurements of changes in the properties of light of a star passing close to the Sun would not be satisfactory for the following reason: since the Sun emits greater amounts of gravitational forces than radiation energies (as evidenced by no appreciable changes in the orbital radii and velocities of the planets since their origin) the shift would naturally be toward red.

In addition to the above test case, we are also proposing two sets of experiments to check the validity of the proposals in these two chapters. The first would be the determination of the velocity of forces emanating from matter and the second the determination of the true wavelengths in different types of redshifts by interferometric methods.

In the determination of the velocities of forces, the easiest method would be the use of either magnetic or coulomb forces. The techniques would be similar to the ones used for the determination of the velocity of light, only in this case instead of using light-sensitive detectors, force-detecting instruments would be used. The direct determinations of the velocities of gravitational forces are much more difficult and probably not practical with the presently available techniques. But there are no reasons to assume that new methods could not be developed for such a purpose.

The interferometric methods for the determination of true wavelengths of lights coming from the distant galaxies need the use of new techniques. The need comes from the fact that two monochromatic lights from two separate sources have to be used in the interferometric measurements: first, the light from a distant galaxy, let us say that of ionized magnesium (Mg II); and second, the light from a source receding at a known speed and emitting light of the same wavelength. This difficulty, however, is compounded by the great possibility that the observed redshift in the lights from distant galaxies is caused by both distance and Doppler effect. In order to differentiate the "mixture redshift" from the other types, two separate determinations have to be made on the same light. The conventional spectrographic determination will give the redshift caused by both. The interferometric method will provide the true redshift caused by Doppler effect. The difference between the two determinations will be the shift caused by distance.

The redshift observed in the light of white dwarfs may be used in checking the possible occurrence of reduction in energies of primatons without an appreciable change in true wavelengths. Since it is claimed here that redshifts caused by gravitational forces are due to the loss of energy of the energy-primatons and not the increase in wavelengths, then the amounts of energy in relation to the true wavelengths in the lights from these dense stars should be less than the amount of energy for the same wavelengths when the measurements are made on lights from standard sources.

★   ★   ★

In conclusion, attention is drawn to the possibility of the revision of the estimates for the size of the observable universe. If it is found that the observed redshifts in the lights from galaxies are due to both Doppler effect and distance, then the distances of these galaxies are shorter than the present calculated values. This in turn would result in proportionately lower luminosities for quasars and the formulation of more tangible explanations of their true natures.

IV

THE NATURE OF EVOLUTIONARY SYSTEMS — THE CREATION OF UNIVERSE

Evolutions and Evolutionary Trends

The word "evolution" may be interpreted in many ways, but generally it is used to characterize a process which exhibits a direction of change or which shows an unfolding, growing, or developing in slow stages that lead toward a definite end. Such an understanding of the word does not include any specifics, but it implies that all systems in states of evolution follow a basic pattern of progression.

Origins and the Progress of Evolutionary Systems

If we follow the modes of origins of the different evolutionary systems, we find that all have more or less the same form: prior to their development, all potential evolutionary systems exist in highly disturbed states in which forces and energies interact chaotically. The formation of the system truly begins if the involved forces and energies become adjusted to each other to the extent that the internal stresses and external interferences cannot break down the temporarily developed state of equilibrium. Once an equilibrium begins to establish itself, the adjustments follow and continue throughout the life of the system, each contributing to the greater stability of the system.

An evolutionary system seldom attains a static state of perfect equilibrium where all the forces and energies are completely neutralizing each other. In almost all cases the system is in a dynamic state, continually approaching the maximum possible stable condition. In some of the simpler systems the state of maximum stability may be achieved, but in the more complex ones the probability of reaching a state of complete equilibrium decreases with the increase in the complexity of the system. *A general description of any evolutionary system may then be given as the course of the formation of any system of interrelated active components progressing toward a state of dynamic equilibrium but seldom attaining a perfectly stable state.* The pattern of the development of such a movement from a system's incipience to its dissolution then becomes its evolutionary trend. In the universe a wide variety of evolutionary systems is encountered, each having peculiarities of its own. But in general all systems follow a basic principle: the greater the number of involved active foci, the longer the time needed for the system to establish its state of equilibrium.

The following examples will serve to illustrate the diversity of the existing systems and some of the concepts of the above statements. The simplest known evolutionary system is that of the hydrogen atom. Its evolutionary progression to a stable state takes only the fraction of a second needed for the electron to be captured by the proton. But the stability of this system is not absolute; under normal conditions the atom is continually being subjected to the outside effects of radiation and gravitational forces. The atom's reactions to such forces prevent it from remaining completely stable; they also permit the atom to be observed and recognized. A hydrogen atom potentially might exist in an absolute stable state if it were located in total darkness, at $0°$ K, and at a point in intergalactic space away from any gravitational influences. There its components would completely neutralize each other's effects and have no interferences from outside. But under such conditions, the atom would give no indications by which its existence could be determined.

A more complex system in equilibrium is the one existing between the bodies of water on Earth and their terrestrial and atmospheric counterparts. Since the cooling of the planet, these components have interacted and formed a recognizable balance among themselves, which is becoming more stabilized as the Earth ages. But the equilibrium is nowhere near completeness. Because of their internal interactions and external influences acting upon them, changes occur continually. Rivers wear away new beds into the land, sea coasts slowly change their shapes, and earthquakes, volcanoes and continental drifts alter topographies. Yet these components exist in a state of equilibrium which for billions of years has been moving toward stability.

Perhaps the most complex example of an evolutionary system is that of life itself. When the conditions on Earth were ready for life, its evolution began and has continued ever since. At present we are witnessing a state of balance between the actively involved units of life, but because of the great number of units and the continual effects of the Sun's radiation on life, this equilibrium exists within a fluctuating system that has very little chance of ever reaching a state of total stability.

Similarly, a star shows an evolutionary trend in its progression. It begins its existence by going through a series of adjustments between the forces and energies of its active foci and comes to establish itself as an entity with a specific identity. Then throughout most of its existence, it demonstrates a state of equilibrium while undergoing many changes in composition and structure. Its maximum state of stability may be considered as when it loses its power to generate energies. Yet even at this point of its evolution, it is still subject to further alterations.

Among the more complex evolutionary systems, any system destined to reach a state of equilibrium begins with a series of disorderly interactions between the involved components. Then as internal forces and energies become adjusted to each other, the disturbances gradually subside. However, these adjustments generally continue throughout the life of a system and principally consist of the gradual neutralization of the forces and energies of the active foci within the system. Thus the *progress* toward a greater stability derives primarily from the process of reducing these energies and forces through their self-neutralization. Since the progress of a system toward stability depends on this process of inter-neutralization, the attained degree of stability of an equilibrium and its response to interferences from outside is controlled mainly by the pattern of the distribution of the active foci within the system.

General Pattern of Evolutionary Trends

An evolutionary system is born when a number of active foci act on each other and form a cohesive entity that remains more or less whole despite the stresses and strains affecting the system from within and without. All evolutionary systems pass through four major evolutionary stages: (1) an origin, when the active foci first come together; (2) a period of adjustments, when additions, eliminations, multiplications, divisions, or shifts in the positions of active foci cause changes in its structure or composition; (3) a long period of stability, in which any great change in its character comes gradually (this phase is usually the longest in duration); and (4) dissolution.

The fact that all evolutionary trends follow approximately the same common pattern in their progress toward stability allows us to draw a general curve indicating the path along which all evolutions advance. The coordinates for this curve would be: the XX' axis representing time and the YY' axis representing the number of the active foci and the rates of their activities in the system. The starting point would be point 0 or the beginning of the formation of the system. This would be followed immediately by the second stage, depicted by a rapidly ascending curve which would then gradually bend toward the horizontal line, where the third stage would begin: the approach to the state of stability and the establishment of equilibrium. The pattern would thus be somewhat similar to the curve shown in Figure 4.

Now if we draw another graph for a mathematical formula related to any phenomenon that can be tested experimentally, such as the relationship between the volume and pressure of a gas, and plot the results obtained within a very wide range of values for x, we find that, if x and f(x) in the formula have a linear relationship, the curve will follow the same general pattern as that in Figure 4; if the relationship is not linear the ascending part of the curve will not be a straight line but the curve will eventually diverge from its regular path and bend toward the horizontal line. In other words, the relationships in nature, as specified by mathematical formulas, hold up within certain limited values, after which a change (at first impercep-

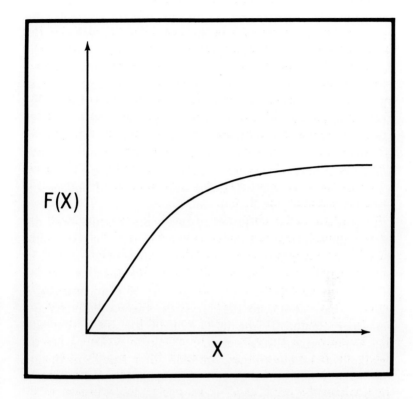

Figure 4. CURVE REPRESENTING THE GENERAL PATTERN OF EVOLU-TIONARY TRENDS. The general trend usually approaches a state of total equilibrium but seldom attains a state of complete stability.

tible but gradually noticeable) takes place in the relationship. As the value of x is increased continually, for the successive increases by the same amount in the value of x, the value of f(x) increases progressively less and less than the predictions of the formula. These responses to the increases in the values of x gradually approach but seldom reach zero. We can then reason that most probably what is true for the relationships that can be checked experimentally is also true for other phenomena in nature. Thus we can conclude: *all natural relationships (unless a change takes place in the physical state of the equilibrium or the relationships are dissolved by interferences from outside) invariably progress toward stable conditions and follow the pattern common to all evolutionary trends.*

One aspect of the evolutionary trends that has not been discussed and which is not shown in the Figure 4 diagram is the pattern of any evolutionary trend's final stage, dissolution. The reason for this avoidance has been the fact that each evolutionary system follows its own pattern of disintegration, a pattern that varies with the effects of other evolutionary systems participating in the dissolution of the system that has reached its terminal stage; usually other systems take part in the breakdown and absorb the by-products into their own. In such dissolving systems, the remnants of the equilibrium disappear as they gradually become parts of other evolutionary systems.

We can now explain the reason for the existence of the phenomenon of "pattern replication." Since the origin and evolution of matter is the prime reason for the existence of all the evolutionary systems, all evolutions are inter-related and share common patterns of progression. Consequently, we should expect replications of patterns among the minor components of different evolutionary systems whenever the conditions in a system are similar to those of another system.

The Creation of Matter

Do the Conservation Laws Interfere with the Concept of Creation of Matter?

The greatest problem in the development of a theory regarding the creation of "something" out of "nothing" is the surmounting of the restrictions of the laws of conservation. This, however, is not as difficult as it may seem because, although we know what "something" is, we do not have any clear notion of what is meant by "nothing." Does the word "nothing," for example, indicate an absolute void or does it describe any state that cannot be perceived by Man or registered by his instruments? The difference is a crucial one for the reason that the laws of conservation deal primarily with energy, force and matter which can be perceived either directly or via devices. What the laws do not take into consideration is the possibility that changes can take place under certain conditions in the no-force, no-energy state of "nothingness," which would result in the state's conversion into energy, force or matter. Similarly, the opposite also holds true; for how can Man perceive the change from what he considers as the state of "existence" to that of "nonexistence" when he does not have any clear idea of where and when to look for such a transformation? The limitations of the laws of conservation are the imposed restrictions resulting from experiments and observations made on Earth, which has had about five billion years to stabilize; this makes the laws limited in their scopes. The problem then narrows down to the consideration of whether or not the scientific laws always remain absolute throughout the universe, and, if not, to establish the extent of their applicability to different phases of the evolution of matter. To do this we need to inquire into the concepts behind the term "scientific law."

A scientific law is a statement formulated from the assessments of the results of numerous experiments and observations for the purpose of predicting the effects that will be produced when specific forces and energies act, either from within or without, on specific systems in equilibrium. Such a law then gives us the means for foretelling fairly accurately the possibilities of the occurrence of phenomena or events. Scientific laws do not take any part in directing the courses of natural events; nor is nature bound or guided by man-made laws, his mathematical formulas or his arbitrary frameworks of reference. The principal function of these laws is to clarify the relationships existing within systems in equilibrium. But since every such system has gone through an evolution of its own and has had to make its beginning from a chaotic state, then *the unvarying and repeatable conditions that have provided the material for the formulation of scientific laws in the immediate past may not have necessarily existed in the distant past.*

We have evidence to show that our scientific laws have not been as immutable in the distant past as they seem to be at present; the best support for this comes from the biological laws. There are many laws related to the living which could not have been formulated prior to the origin of life nor even a considerable time after the onset of that major evolutionary system. This is the reason: during the initial period in the evolution of life the states of equilibria within the individual units, among the evolving units, and between the units and the environment had not become stable enough to provide the repeatable and unvarying responses that are needed for the formulation of laws. Since the condition of stability cannot be met in the initial stages of any of the major evolutionary systems, the prediction capabilities of the known laws are nil prior to the origin—and very limited during the first stages of all evolutionary systems. Consequently, *the degree of predictability and the applicability of any scientific law goes through an evolution of its own, evolving step by step with the evolutionary system to which it applies.*

The laws of conservation have been formulated within the stage of maximum stability in the evolution of matter and as such are only applicable to that phase of that evolutionary system. The experiments and observations that support these laws have been performed on Earth, a planet which has had about five billion years to stabilize. Men have used equipments and instruments made of earthly material in order to prove that matter, energy, force, etc. can neither be created nor destroyed. But as these investigations do not include material from the early stages of the evolution of matter, they cannot give correct assessments of whether or not they also apply to matter in the period prior to the formation of Earth. Therefore, *there are no proofs or evidences to suggest that the laws of conservation are also applicable to the initial stages of the evolution of matter; hence they can in no way oppose a concept of the creation of matter from the state of "nothingness."*

The Origin of Matter

In this book the state of non-uniform motions of "primatons with independent motions" described in Chapter II is regarded as the state of "nothingness"; this is as much as we can presently explain; it would be futile even to try to contemplate how or why primatons originated. We know that the orderliness of motions is the fundamental requirement for the existence of matter. Therefore, we shall consider the creation of matter as the transformation of the disorderly state of "independent motions" of primatons to the state of orderly "influenced motions."

Before we proceed any further, note should be taken about the usage of certain terms. One of the problems of writing about the origin of matter is the use of proper words to describe and characterize the hitherto unconsidered states now proposed. When one talks about the origin of matter, one has to go very far back to the beginning of what we call "time"; consequently, such terms as "force," "energy," "era," "period," "when," "where," "before," "after," etc., become meaningless. These words were made to express post-creation conditions. Yet unless new terms are forged, a cumbersome and complicated undertaking, one is forced to apply old expressions and terms in an inaccurate and misleading way. Hopefully the reader will remember this point and, we, on our part, shall try to avoid the misuse of words as much as possible.

In 1965, Hannes Alfvén* proposed a theory of the origin of matter based on the concepts originally described by O. Klein, where the formation of symmetrical particles from energy is used as a foundation for the creation of matter. We shall attempt to follow this pattern of thought and modify it by the application of some of the hypotheses presented in this book.

The prime requirement for the production of a pair of symmetrical particles is the introduction of sufficient amounts of energy into a field of force. Energy by itself does not produce particles. To have a field of force we need the existence of

*Hannes Alfvén: "Antimatter and the Development of the Metagalaxy," *Review of Modern Physics*, 37: 652 (1965). A review of this article is given by the same author in *Worlds-Antiworlds*, W.H. Freeman and Co., San Francisco, California (1966).

particles and the uniformity of motions of primatons around them. This means that the creation of matter probably went through two stages: (1) the formation of the original particles somewhere in space; and (2) the self-breeding of these particles.

It is proposed that the first stage came into being somewhere in space where, according to the provisions of the law of probability, an abnormal congestion of the primatons with "independent motions" occurred—let us say within a sphere of about 10,000 light years in diameter. We shall refer to this region as the "congested sphere." The high concentration of the primatons in the sphere caused the gradual conversion of their disorderly motions to orderly ones via a large number of collisions. Our main concern now will be to show how the original stable particles and orderly moving primatons came into existence from the chaotic state of the disorderly primatons within the "congested sphere."

The unique property of primatons is their absolute rigidity: they do not bounce off each other when they collide. Two types of collisions are possible for primatons: either they collide along a line connecting the two centers or they glance off each other if their paths form an angle. In the former case, the two adhere and in the latter case the two separate, each gaining a rotary motion at the expense of part of its linear motion. Because of the spherical shape of the primatons, a single primaton experiences many glancing collisions before it encounters an adhering one.

When two single primatons glance, the direction of their new linear motions is a function of the angle between their paths prior to contact. Also, each acquires a rotation along an axis perpendicular to the new linear motion, with the directions of rotations of the two opposite each other. Many such collisions would have caused the relative motions of the disorderly primatons to become more uniform; the greater the number of collisions the greater the uniformity.

Occasionally a collision would have occurred along the line connecting the centers of two primatons and the two would have adhered. The newly formed doublet would have provided a greater surface of contact with reduced chances of glancing on collision with a third primaton. With increased opportunities for collisions as the unit became larger, the rate of increase in size would have improved and eventually resulted in a spherical particle. Many such spheroidal particles came into being; then disintegrated when they collided with other spheroids or very fast-moving primatons. But each collision and union improved the general motions of all primatons in the region toward greater uniformity.

Eventually, one stable particle (either proton or antiproton) came into being. This was the nucleus from which all the fundamental particles were produced. We know that, for a pair production to occur, an amount of energy equivalent to the mass of the two particles has to be introduced into a field of force. Presently, there is a maximum limit for the speed of primatons; hence there is a maximum amount of energy a primaton can carry, and this is not sufficient to cause proton-antiproton pair production. But in the "congested sphere" there was no such speed limit; consequently some of the primatons entered the field around the particle with speeds far greater than the speed of light. Although the velocities of most of the primatons inside the "congested sphere" had become somewhat uniform, there was a continuous influx of the "independent primatons" with some of them moving at extremely fast rates relative to the particle. It has been stipulated that the linear velocity of a primaton determines its energy; therefore some of the primatons with "independent motions" could have the required energy needed for pair production. It is believed that a number of pair productions occurred within the "congested sphere" and as a result a number of protons and antiprotons were produced from the original particle.

Because some of these particles were also moving at very high speeds relative to one another, every so often there was a collision between two particles, producing a number of intermediate particles which disintegrated into energy, neutrinos, electrons and positrons. This was probably the source for the two smaller fundamental particles. The four stable elementary particles had come into being.

The creation of matter could be divided into two main stages with the first gradually phasing into the second. The first stage would be the creation of original particles and the establishment of uniformity of motions (such as the limit of linear speed to that of velocity of light) among the primatons inside the "congested sphere." The second would comprise the self-breeding of the particles and their effects on the free primatons. The pair production was a self-perpetuating process that caused the formation of particles but also assisted in the transformation of the disorderly motions of free primatons to those of force and energy. As a result, the linear speeds of most of the primatons around the created matter became uniform and a balance became established between the motions of the free primatons with those of the particles. This was the real origin of matter; what followed brought into existence the universe made of matter.

The final outcome of all the transformations was the formation of countless $+H$ and *antiH* atoms. Alfvén has pointed out that a mixture of these antagonistic atoms would not necessarily result in complete annihilation. There is more likelihood that the two types of matter would remain separate through the effects of the Leidenfrost phenomenon. This means the formation of alternate regions in space where one type of matter would be dominant in one region. Once the dominance of one type became established, that type prevailed and the newly created antagonistic atoms in that region were either annihilated or driven away. In this way total segregation of the two types of matter took place, resulting in regions consisting exclusively either of $+H$ or *antiH*.

As a consequence of his hypothesis of the creation of equal numbers of $+matter$ and *antimatter* atoms, Alfvén suggested four possibilities about the existence of *antimatter* in the universe. In his fourth supposition he has proposed cautiously the probability that every second star in our vicinity is made of *antimatter*. We agree with this opinion and go a step further. It is our belief that not only is there an even distribution of $+matter$ and *antimatter* stars in the universe, but that even the existence of individual stars and planets is dependent on the balance that exists between $+matter$ and *antimatter* in each star. This modification necessitates a basic change in the interpretation of the data at hand. The rest of this book is devoted to this task.

A General Scheme for the Formation of Stars

The creation of matter was a slow and gradual process. Consequently the heat generated by $+H$ and *antiH* encounters, or other causes, had no chance to accumulate and dissipated into space; this left the region at about $0°$ K for a long time. The result was the freezing of the newly created atoms into solid masses. In the beginning the masses were small but gradually increased in size as they attracted newly formed atoms and molecules of their own type. As a result, frozen masses of various sizes appeared in each region, some of gigantic magnitudes. The enlargement of the mass could not produce any compression at the center, as at $0°$ K, matter is at its maximum compressed state. Any generated heat inside the mass would have been disseminated into space; with the entire mass being in a superconductive state at about $0°$ K, heat would have been transferred to the surface at a rapid rate and radiated into space. Also, because the rate of growth of the masses was very slow, any generated heat would have had plenty of time to leave the mass. Consequently, the masses could grow into any possible size and remain in the frozen state. The only limiting factor was the availability of material.

The usual consequence of the continual accumulation of the newly created atoms was the formation of a very large frozen central mass around which smaller masses of the same type orbited. We shall discuss the formation of these systems in detail in the next chapter. It will be seen that in most cases only one such system would come into being within each region. Up to this stage of their evolution, the frozen systems could not move out of their own regions, but as the conversion of the disorderly primatons progressed and the resulting particles became incorporated into the frozen masses, the attraction between the neighboring systems began to increase. The large frozen masses and their satellites would then move toward each other.

Whenever the mutual gravitational attraction between two adjacent systems became effective enough, the two moved toward one another. If two systems of the same type reached each other, they joined and formed an even larger system. If the two systems were of opposite types, the encounter became full of numerous incidents, which in most cases resulted in the formation of a star.

The details of the formation of a star will be given in the next chapter when we discuss the evolution of the solar system. At this juncture we shall provide only a general sketch. The first step toward the formation of a star was the coming together of two antagonistic systems—one made of *antimatter*, the other made of $+matter$ and each composed of numerous satellites orbiting around a very large central mass. The resulting entanglement brought about a number of $+matter$-*antimatter* *reactions*. The radiation from these sources evaporated other satellites causing the mixing up of $+H$ and *antiH* gases and radiation of considerable amounts of energy. The largest mass, however, continued to remain frozen for some time, with the evaporated gases that were surrounding it, insulating it from outside heat; in return, the central frozen mass held the shielding gases by its gravitational attraction. The picture at this time was that of a large central mass, frozen in the center and surrounded by an atmosphere of homogeneous gases which in turn were surrounded by a shell composed of a mixture of $+H$ and *antiH* gases.

Gradually the two types of gases in the outer atmosphere separated; the homogeneous type joined the atmosphere immediately above the frozen mass; and the antagonistic atoms remained above the main atmosphere, keeping at a distance from it by a radiation zone that had appeared between the two. From then on there was a continuous annihilation of the atoms that penetrated and passed through the radiation zone, with the process generating a continuous and relatively uniform amount of radiation. This caused the gradual heating of the central body; eventually it became totally gaseous and incandescent. The process of the formation of the star was by then complete. Its structure was composed of a large spherical mass of luminous hot gases surrounded by a tenuous antagonistic atmosphere. The atmosphere was held by the great gravitational forces from the central mass. The concentration of the atoms in the atmosphere was a result of the balance between the generated energy and the gravitational forces. Since the atmosphere was not confined on the outside, it extended far into space, becoming more tenous with distance.

Special Effects Produced during the Formation and Evolution of Stars

The *+matter-antimatter reactions* are most probably responsible for the appearance of the following phenomena:

1. The encounters between *+matter* and *antimatter* atoms are always accompanied with the generation of electromagnetic waves, but the type of encounter determines the dominance of a certain range of wavelengths. In the milder reactions, such as the ones that occur in certain nebulae (details in Chapter VII), the velocities of collision are not great enough to overcome the repulsive energies produced at the point of contact between the two antagonistic atoms; after their encounter the two antagonistic atoms separate. The radiations from these sources are mild and composed mostly of radiowaves. In contrast, whenever the conditions of encounters result in total annihilation, the emissions from these sources contain a large percentage of high-energy shortwave radiations.

2. The *+H-antiH reactions* resulting in the total annihilation of atoms cause the formation of higher atomic weight elements from the nearby atoms. Neutrons are first produced from the hydrogen atoms (or antineutrons from antihydrogen atoms); the powerful energies drive the electrons into their respective protons to form neutrons. The neutrons are then captured by other hydrogen atoms to form hydrogen and helium isotopes. The high temperatures then bring about the combination of these nuclei with the elements present to produce heavier ones. The details of these transformations are already available and may be found in the literature.

It is believed that the establishment of the equilibrium of a star is always accompanied by the appearance of a significant phenomenon which characterizes the unit throughout its life and plays a very important role in its evolution as well as the evolution of the galaxy to which it belongs. The phenomenon is as follows: the gravitational constant of a star increases very gradually as it grows older, the rate of this increase being a function of the rate of energy produced. The explanation of this phenomenon is based on the principle that the increase in the mass of a moving atom inside a space confined by a radiation zone is cumulative. True, the velocities of atoms inside stars are not great enough to produce noticeable increases in their masses within a short period of time, but billions of years of continuous accumulation can produce marked increases in the strengths of emitted force-primatons. The amount of this increase is equal to the amount of energy received and converted to mass by the atoms since the time they became incorporated in the body of the star. In turn, the increase in the mass of a particle is accompanied by an increase in volume; this results, not in appreciably greater numbers of gravitational force-primatons being emitted, but in about the same numbers having relatively faster rates of rotation and hence stronger force per primaton. *Accordingly, the present-day strength of gravitational forces emitted by any star is the product of two factors: its mass and the amount of energy it has produced during its life.*

As a star ages, the increase in its gravitational constant causes an increase in the density of its atoms and hence in its gravitational attraction. Here is an apparent paradox: the annihilation of part of the mass of the star causes it to have a greater attractive power. However, we believe that by the application of this phenomenon we can describe the evolutionary courses of stars and galaxies.

It might be argued that we do not have any evidence for the existence of variability in the masses of the stable particles. The answer is that the lack of direct evidence does not necessarily negate the premise, especially when one considers that the increase in the masses of the atoms occurs inside the stars with little chance for them to escape and reach the Earth. However, we probably do have indirect support for this phenomenon from two unrelated sources. First, we have the "continuous spectra" from the atmospheres of some of the white dwarfs. As we shall see later, the atmospheres of these dwarfs may be explained in terms of dense gases that have come from the interiors of old stars, and since the electrons of the dense atoms do not oscillate in the manner of the normal ones they cannot emit specific spectral lines. Second, the possibility exists that some of the cosmic-ray particles have masses greater than the assumed values, especially the ones with calculated relativistic velocities. The determination of the velocities of these particles has been based on the assumption that the identified particle has a fixed and known mass and that the computed momentum is entirely due to the velocity. But it is conceivable that a part of the value calculated as velocity belongs to a greater than expected mass for the particle.

Finally, it should be mentioned that since the atmospheres of stars (except white dwarfs and the ones in planetary nebulae) are not confined, there is almost no possibility for an increase in the value of their gravitational constant.

General Schemes for the Formation of Galaxies and the Origin of the Universe

The details of the evolution of galaxies will be given in Chapter IX. The following resume of the major evolutionary stages is for the purpose of laying the foundation for the coming discussions on the related evolutions of the solar systems, stars and other observed phenomena in the Galaxy. Notice should be taken of the fact that in this hypothesis the universe is believed to be much older than the presently computed values of about twenty billion years.

Formation of Galaxies

The following is the tabulation of the major stages in the origin and the evolution of galaxies:

1. The process of pair formation and the consequent segregation of *+matter* and *antimatter* into exclusive regions spread out radially from the point of origin into space. The first star made its appearance either inside or in the vicinity of the "congested sphere."

2. The initial few hundred stars came into being in the neighborhood of the first star. As they aged and their mutual at-

tractions increased, they moved closer to one another to form a semblance of a globular cluster.

3. As matter became available, new stars came into being at greater and greater distances from the center, but the continual increases in the gravitational constants of the central cluster and the aging stars drew them closer to form the embryo of a galactic nucleus.

4. New stars continued to appear farther and farther away from the center, but their motions were under the influence of the combined gravitational forces of the stars in the original nucleus. Since the strengths of the attractive forces between the stars and the central structure were increasing with age, new stars were being pulled toward the center, and this continually increased the size and the gravitational attraction of the nucleus. The structure of the nucleus at this time was that of a compact center, surrounded by stars whose number per unit of space volume diminished with distance; this means that there was no boundary separating the central structure from the rest.

5. Meanwhile the process of star formation was spreading radially into space; the farther the star from the center the less influence the nucleus had on its motions.

6. The form of the galaxy was determined by the pattern of motions of the stars that were within the influence of the nucleus and the mode of their accumulation around the central structure.

7. While this formation was taking place, other stars were coming into existence and eventually a stage was reached when distance became an effective factor; beyond a certain radius the nucleus and its satellite stars could not control the motions of the newly formed stars. The motions of these independent stars could be either toward or away from the nucleus; the ones with the former type of motion eventually joined the galaxy and the ones with the latter type of motion pulled away to become members of the second generation galaxies.

8. While moving away, these stars became divided into groups, each group in the form of a galactic cluster which later converged; the convergence was caused by the con-

tinual increase in their mutual attractions as the stars grew older. The total picture at this stage was that of a spherical region containing an evolving galaxy surrounded by a number of galactic-type clusters moving away radially from the central region.

9. In the course of their motions the stars of the galactic clusters gradually moved close to each other, through their increased mutual attraction, to become globular-type clusters. Some of the clusters became centers for attracting the newly formed stars in their regions and gradually evolved into nuclei of second generation galaxies.

10. In turn the second generation galaxies gave birth to third generation galaxies. Then generation followed generation to produce the universe. The pattern of the formation of galaxies up to the fifth generation is shown diagramatically in Figure 5, but the movements should be envisioned as expansions in three dimensions instead of the two dimensions shown in the diagram.

Pattern of Expansion of the Universe

With regard to the general scheme for the evolution of the galaxies and the formation of the universe, attention should be drawn to the following:

1. In the process of its formation, each galaxy drew its material from its surroundings, a region roughly spherical in shape. In our coming discussions we shall refer to this region as the "galactic supply region."

2. At present we cannot assign any specific age to the universe, but an estimate of an approximate age may be worked out after the determination of the position of the point of origin of the universe.

3. This hypothesis places a limit to the size of the "matter universe," an expanding region within an infinite space containing primatons with "independent motions." However, since there are no reasons to believe that the conversion of the disorderly primatons to matter has stopped or will ever come to a halt, the present limit may be considered as temporary. More and more stars are coming into existence at the bound-

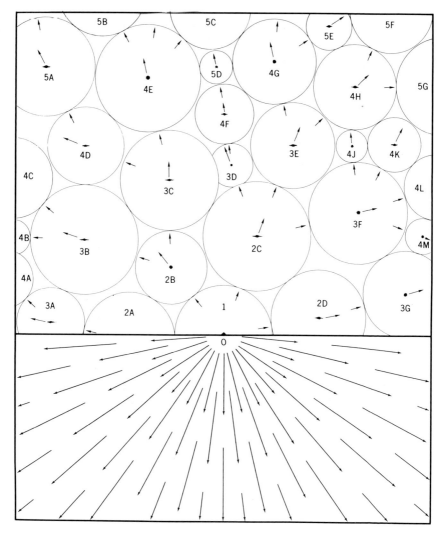

Figure 5. DIAGRAM SHOWING THE MANNER OF THE FORMATION OF GALAXIES FROM THE ORIGINAL GALAXY. This represents a cross section of a three-dimensional region. (1) The top frame demonstrates the pattern of births of galaxies: numbers indicate generations; letters identify members belonging to a generation; an arrow next to any galaxy determines the direction of its motion relative to the point of origin; the circles represent the approximate "galactic supply regions"; the arrows near the periphery of any "supply region" indicate the directions of motions of the galactic-type clusters originating from that region. (2) The bottom frame shows the general pattern for the transformation of disorderly primatons to matter. This is also the basic pattern for the expansion of the universe.

ary between the orderly and disorderly regions in space, and this is continually increasing the size of the universe composed of matter.

4. The boundary between the "matter universe" and the space beyond may be pictured as a shell-like spherical region where the creation of matter is taking place. The formation of the new matter and the births of stars is associated with great amounts of $+H$-$antiH$ *reactions* and continuous generation of electromagnetic waves. At this stage the matter is in a diffused form and consequently there are no isolated strong sources of radiation. A part of the emissions from this region propagates inward, but since this radiation has to travel long distances to reach us, a large portion of its energy is lost to space (redshift caused by distance). Thus by the time we intercept these electromagnetic waves they are weak and more or less homogeneous. *We may then consider the observed isotropic cosmic background radiation as a strong indication of continuous creation of matter at the boundary between the "matter universe" and the "no-matter space."*

The "matter universe" is encroaching into the space containing disorderly primatons in another way, through expansion. The rate of this expansion, however, is not as rapid as the calculations from redshift determinations indicate. It would be reasonable to assume that the rate of motion of any newly formed galaxy relative to its mother galaxy would have been of the same order as the average of the present rates of motions of globular clusters relative to the nucleus of the Milky Way. Therefore the rate of recession of any galaxy from the point of origin of the universe would be dependent on the number of times its generation is removed from the original galaxy; for each generation the galaxy's rate of recession would be greater by a factor equal to the average of the rates of motions of the globular clusters. This is a picture of an expanding universe, but quite different from the one envisioned by the prevalent theory. There are three principal differences between the two concepts:

1. In our scheme the rates of recession are slower than the calculated rates based on interpretations exclusively in terms of Doppler effect.

2. As shown in the Figure 5 diagram, in our hypothesis the directions of movements of galaxies do not have to be entirely radial. Consequently, some of the galaxies can actually move toward each other while receding from the point of origin. This is shown in the cases of galaxies 3E and 4J in the diagram. Most probably the motion of Andromeda toward Milky Way is this type of motion.

3. According to our interpretations there is a point of origin in space. The corrected rates of motions of galaxies according to the suggestions in the previous chapter should direct us to that position. The galaxies near this point should show the properties of old galaxies (see Chapter IX regarding the identification of older galaxies). There are no such provisions in the prevalent theory of expanding universe.

In conclusion, notice should be taken of the fact that the main theme of this book, "the phenomenon of pattern replication," has been very much in evidence in this chapter. The pattern of the creation of matter, the attraction between +*matter* and *antimatter* to produce a star, the aggregation of a number of stars to form colonies in the shape of galaxies, the birth of new generations of galaxies from the older ones, all have their counterparts in the evolution of life. But, as we have already mentioned, it is life that has replicated the pattern of evolution of matter and not vice versa. There is only one fundamental pattern of evolution in the universe; all others are copies of that master-pattern.

V

THE EVOLUTION OF
THE SOLAR SYSTEM

The solar system, the best studied and the most observed system in the universe, provides us with valuable clues regarding the origin, the structure and the evolution of the majority of stars. We have already described in a general way a new hypothesis for the stellar origins. Here, we shall elaborate on the previous statements by describing the three main stages of the evolution of the solar system. As we progress we expect to provide explanations for the following characteristics of the system:

1. The properties of the interior of the Sun and its atmosphere.
2. The reason for the manner of distribution of the angular momentum in the solar system.
3. The causes for the patterns of motions of most members of the system.
4. The manner of development of some of the pertinent properties of the planets and their satellites.
5. The properties and nature of comets.

The First Evolutionary Stage —
The Origin

We shall first outline a general scheme for the formation of the frozen primordial systems which we believe to have been the precursors of all stars and planets. Then we shall describe the specifics in the structures and the manner of unification of the two systems which resulted in the origin of the solar system. In the next chapter we shall elaborate on the effects of the characteristics of the primordial frozen systems in the production of the stars of the H-R diagram.

The history of the solar system begins in two adjacent regions within the incipient "Orion arm" of the Galaxy, which at that time were located much farther away from the nucleus than the present distance of the Sun from the center of the Milky Way. Each of the regions supplied its own type of matter; one provided $+H$ atoms which after accumulation into numerous frozen masses evolved into the smaller "primordial Jupiter system," and the other provided $antiH$ atoms which resulted in the formation of the much larger "primordial Sun system" (containing more than 95% of the total mass of the two systems).

Formation of the Primordial Frozen Systems

In the early stages, the created matter segregated into $+H$ and $antiH$ regions. Within each of these regions a number of frozen masses were formed. Each mass then accumulated atoms from its surrounding. Depending on the order of the appearance of the frozen masses, their proximity to each other, and their respective positions within the region, they increased in size, but in the majority of cases one mass grew at a faster rate than the rest and gradually gained dominance over the whole population of the small frozen masses. This dominating mass evolved into a primordial frozen system.

The mass that was destined to become the central body of the primordial frozen system was a spinning sphere, gradually increasing in size and continually attracting atoms from every possible direction. Since the centrifugal forces in a rotating sphere are at their greatest strengths at the equatorial region, the increase in the size of a frozen mass meant greater and greater centrifugal forces acting at its equator. Eventually these forces caused the breakage of a comparatively large part at a weak location, and this piece spiralled away along the equatorial plane of the frozen sphere. Also, since at the time of the breakage the outer side of the broken piece had a greater angular momentum (greater distance from the center) than the inner side, the departing mass developed a spin in the same direction as the direction of rotation of the central mass. The breakage, in turn, caused a corresponding slowing down of the rate of rotation of the central body. This meant that the mass could grow larger before the forces could again become great enough to cause a second significant breakage. Since the rate of accumulation of matter was fairly slow, the second separation occurred a considerable time after the first piece had departed and had moved some distance away. In between the major separations there were also numerous small breakages. A succession of such breakages and slowdowns of rates of rotations resulted in the formation of a system composed of a comparatively large rotating central body surrounded by a multitude of satellites of different sizes, all spiralling away from the center along the equatorial plane of the central body and all rotating in the same direction as the direction of the spin of their mother-mass. Most of these breakages took place when the central mass was still fairly small.

While the atoms were being attracted and added continually to the central mass (much greater attractive forces), little material could be gathered by the much smaller spiralling satellites as long as they were near the center. But as their distances from the mother-mass grew greater, they in turn reached regions with ample supplies and began to increase in size. The continued increase in the masses of the larger satellites as well as that of the central body gradually caused increased gravitational attractions between the central mass and its satellites; this in turn caused proportional reductions in the rates of recession from the center. Eventually, one by one the larger satellites could no longer continue with their away movements, and each went into an orbit of its own around the central mass. During this period the largest satellites went through the same cycles, each producing a number of secondary satellites orbiting along its equatorial plane, and all being coplanar with the equatorial plane of the central mass and spinning in the same direction.

The amount of the available material which was principally a function of the size of the supply region was the fundamental factor in determining the extent of the progress of the evolution of these primordial frozen systems. This may be demonstrated by the following three examples:

1. If the size of the supply region was relatively small, the evolution of the system would have reached its stage of completion shortly after the large satellites went into orbits. Those

that did go into orbit would have remained in the system, and the ones that could not be held by the central mass would have spiralled away into space and become lost. This is believed to be the way the primordial Jupiter system (Figure 6) came into being.

2. If the supply region was of medium size and a considerable amount of material was still available after most of the larger satellites had gone into orbit, the central body and the larger satellites would have continued to grow in size; the increases in the masses would have resulted in the continual reduction of the orbital radii caused by greater gravitational attractions. Since the degree of these reductions depended on the availability of material, in a medium-sized supply region the contraction process came to a halt before many of the larger satellites had come too close to the central body. At this time the system consisted of a relatively large central body with a considerable number of larger satellites orbiting near the central mass. The total mass of this system was much greater than the one mentioned in the first case, but because of the contraction of orbits the diameter of the system was about the same or even smaller than the one belonging to the less massive system mentioned above. It is suggested this was the manner of evolution of the primordial Sun system (Figure 7).

3. In the case of a large supply region, most of the larger satellites would have been pulled in very close to the central mass, resulting in a system composed of a very massive central body surrounded by many large satellites orbiting very close to the center. The diameter of this system at its equatorial plane would have been smaller than the previously mentioned systems. This type of system probably would have evolved into a massive star.

The above are merely three examples selected arbitrarily from a large variety of frozen primordial systems. As we shall see in the coming chapters, the size and mass of the primordial frozen systems have had great influences in the formation of the stars of the H-R diagram.

Figure 6. DIAGRAM OF THE SUGGESTED MODE OF DISTRIBUTION OF THE SATELLITES OF THE FROZEN PRIMORDIAL JUPITER SYSTEM. Only a few of the larger satellites are in the vicinity of the central body.

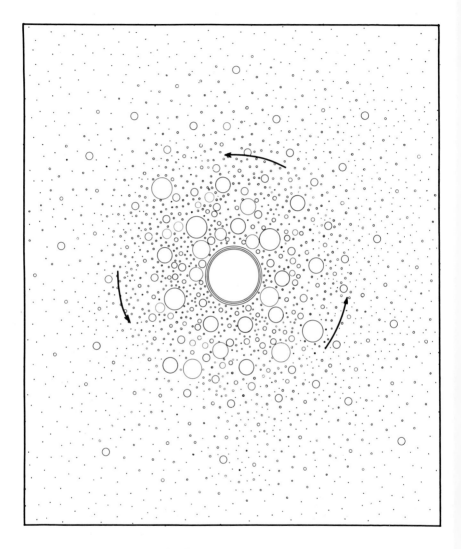

Figure 7. DIAGRAM OF THE PROPOSED MODE OF DISTRIBUTION OF THE SATELLITES OF THE FROZEN PRIMORDIAL SUN SYSTEM. Most of the larger satellites have been pulled in and are orbiting quite close to the central body.

Evolution of the Primordial Jupiter System

We have already given a general scheme for the development of the frozen primordial systems. Now we shall describe more specifically the manner of formation of the satellites of the primordial Jupiter, with two points in mind: first, the pattern of the orbital motions of the larger satellites around the primordial Jupiter should account for the present pattern of orbital motions of the planets around the Sun—that is, the nearer the planet is to the central body the faster is its orbital velocity; and second, the present motions of the original satellites of Earth, Mars, Jupiter, Saturn and Neptune (which have the same pattern of orbital motions as the planets) should have originated during the evolution of the primordial Jupiter system. As we shall see shortly, the present planetary orbital motions are the outcome of two main factors, the pattern of the orbital motions of the larger frozen satellites of the primordial Jupiter, and the positions of these satellites relative to their central body at the time of the capture of the primordial Jupiter system by the primordial Sun system. It will also be shown that the present pattern of motions of the Jovian satellites is the product of the evolution of the primordial Jupiter system.

We shall begin with the period when the away motions of the larger satellites in the primordial Jupiter system had come to a halt and all had gone into orbit around their mother-mass. In addition to the very large ones, there were five medium-sized satellites orbiting relatively close to the central body. These five, being the last of the large breakages, could not accumulate matter at the same rate as their mother-mass and therefore had gone into orbit for the principal reason that shortly after their departure the mass of the primordial Jupiter grew large enough to arrest their away motions. However, the orbital radii of all these medium-sized satellites at that time were considerably larger than the present orbital distances of Jupiter's inner moons; this was also true for the distances between the orbits of the larger satellites and the primordial Jupiter as well as the orbital distances of the secondary satellites relative to their respective mother-masses.

At this period all the available material had not as yet become exhausted, and further accumulation of the remaining atoms brought about the contraction of the system. Two interdependent phenomena were taking place at this time: the gravitational attractions between the orbiting satellites and their respective mother-masses were increasing, and this in turn was causing the shortening of all the orbital radii. The angular momentum of an orbiting body is the product of its mass, its velocity and the radius of its orbit. Consequently, in order to conserve the angular momentum, the orbital velocities of the satellites increased as their orbiting radii became smaller. But the increase in the attraction was due principally to the increase in the mass of the central body, a factor which influenced all the satellites; the central mass was much larger, attracted greater number of atoms and grew in size at a much faster rate. According to the provisions of the law of the inverse square of distance, for any specific amount of increase in attraction, the nearer a satellite was to the center the greater was the reduction of its distance from the central body and consequently the faster its final orbital velocity.

So far we have spoken of the large and medium-sized satellites. We shall now describe the role of the myriads of small frozen masses that were the natural products of the minor breakages. Most of these were spiralling away in the equatorial plane. As long as they were within the effective ranges of the larger satellites, they could accumulate only meager amounts of material, but as they moved away from that region in the system they received enough material to make them become more responsive to the gravitational pull of the primordial Jupiter. The increased attraction eventually brought their away movements to a halt; the largest ones were the first and the smallest were the last to go into orbit. However, this pattern of distribution was not uniform, but as a rule the smaller the mass the farther it was from the center. By the time the system came into equilibrium, a large number of small, frozen satellites were orbiting outside the orbit of the farthest large satellite. As we shall see shortly, in the encounter between the two antagonistic systems these masses played an important part in initiating the destruction of the angular momentum of the primordial Sun system.

To summarize, the general picture of the primordial Jupiter system before its capture was as follows:

1. The central body was a relatively large, spinning, spherical frozen mass composed of $+H$, somewhat larger than the present Jupiter (about 1-1/2 to 2 times as large—part of the mass was lost during and after the capture), surrounded by at least ten large orbiting satellites (the masses varying between those of present Saturn and Uranus), interdispersed with a considerable number of medium-sized and small frozen masses of various sizes. The orbital velocities of the large satellites followed the general pattern: the farther away from the center the slower the velocity. The orbital velocities of the smaller ones did not follow this rule.

2. Five medium-sized satellites, much larger than the present five Jovian inner moons, were orbiting close to the primordial Jupiter on the equatorial plane of the central mass.

3. Each of the large satellites had a number of secondary satellites of its own, the numbers and sizes depending on the amount of material they had been able to accumulate as they moved along their orbits. The secondary satellites that had descended directly orbited along the equatorial plane; the nearer the satellite was to its central body, the greater was its orbital velocity; the direction of their rotations was the same as that of their mother-masses.

4. Countless numbers of small frozen masses orbited on the primordial Jupiter's equatorial plane outside the orbit of the farthest large satellite. As a rule, the farther the mass was from the center, the smaller was the size, but there were many exceptions to this general pattern. The rule applicable to the orbital velocities of the large satellites mentioned above did not apply to this group; the small masses were moving at velocities comparable to their own velocities at the time of their breakage from the central mass.

5. In addition to the small satellites orbiting on the primordial Jupiter's equatorial plane, there were a number of frozen masses, varying in sizes, orbiting at any random orbit and direction around the primordial Jupiter and some of its larger satellites. These were the captured frozen masses that came into being with Jupiter and which had not been able to

accumulate large amounts of matter. Consideration should be given to the possibility that some of the contemporary satellites of the outer planets with abnormal orbits are remnants of systems (miniature forms of the primordial Jupiter system) that were captured by the larger frozen satellites in the same way the primordial Jupiter system was captured later by the primordial Sun system. It should also be remembered that many of the independent satellites were destroyed during the capture.

Evolution of the Primordial Sun System

Unlike the primordial Jupiter system, there are very few clues regarding the mode of development of the primordial Sun system. However, we do know that the young primordial Sun rotated at speeds comparable to that of the growing primordial Jupiter, for otherwise it would not have possessed the satellites that brought about the drastic changes in the compositions and masses of some of the satellites of the primordial Jupiter system during and after the event of capture. Therefore, of necessity we make the assumption that in the early stages of its formation the rate of rotation of the primordial Sun was about the same as that of the young Jupiter. The pattern of the evolution of both would have been more or less the same. The cause of the differences in the final structure of the two systems was most probably the availability of material. The primordial Sun system received far greater quantities, and this gave it the typical characteristics we have outlined for the group of frozen primordial systems with medium-sized supply regions. This brings out the possibility of the final rate of rotation of the primordial Sun having been slower than that of the primordial Jupiter.

Most probably the final mass of the primordial Sun was not as large as the present mass of the Sun. The present mass is the result of the annexation of the contents of most of the primordial Sun's satellites; in this estimate we do not include the losses due to transformation of mass to energy and the increase in gravitational constant caused by aging. A central body of that size should have had a number of large satellites, some of them even larger than the primordial Jupiter. This means that at the time of capture most of the satellites were orbiting close to the center. The majority of the large satellites could not have been much farther than the present orbit of Mars, with most of the congestion probably inside the present orbit of Venus. Although the number of the larger satellites became reduced as the distance from the center increased, similar to the primordial Jupiter system, myriads of smaller satellites orbited on the primordial Sun's equatorial plane all the way to the present orbit of Neptune, and a limited number orbited even beyond that region. However, as we shall see presently, the relative distances of the terrestrial planets from the Sun and the manner of distribution of the densities of the contemporary inner moons of Jupiter point to a pronounced reduction in the size and number of the frozen *antiH* satellites beyond the present orbit of Jupiter.

The Capture

The two antagonistic systems that had evolved separately began to move toward each other through the effects of increases in their mutual gravitational attraction. The much smaller one, the primordial Jupiter system, was composed of $+H$ and the primordial Sun system composed of *antiH*. The two systems became attracted to each other while still in their growing stages, with the maximum rate occurring after most of the gases had been accumulated.

Figure 8 is a schematic representation of the proposed manner of the capture of the primordial Jupiter system by the primordial Sun system. For the sake of simplicity, in the diagram only three of the primordial masses belonging to the captured system are used: the primordial Earth and Uranus positioned approximately opposite each other relative to the primordial Jupiter. The uncertainties pertaining to the relative positions of the satellites while their very slow capture was taking place as well as the effects of the $+H$-*antiH* reactions that followed, make it extremely difficult to plot the true courses of the movements of the primordial Jupiter's satellites during this period. The diagram is supposed to convey a very simplified pattern of

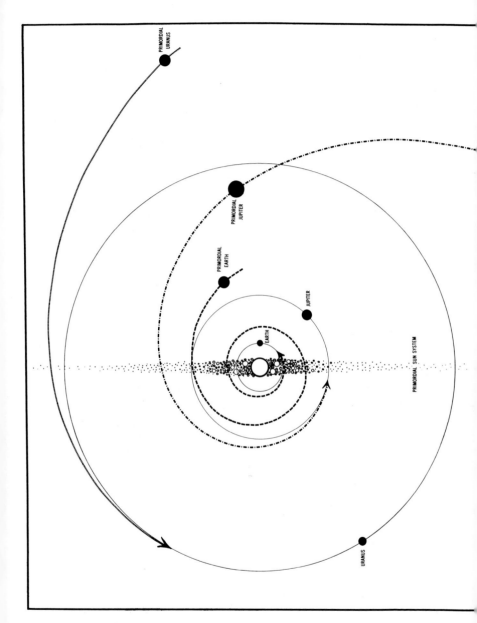

Figure 8. **SIMPLIFIED DIAGRAMMATIC REPRESENTATION OF THE CAPTURE OF THE PRIMORDIAL JUPITER SYSTEM BY THE PRIMORDIAL SUN SYSTEM.** The primordial Uranus represents the large frozen satellites of the primordial Jupiter system that evolved into the outer planets. The primordial Earth represents the large frozen satellites that evolved into the terrestrial planets.

the capture, Uranus representing the satellites that eventually became the outer planets, and the Earth depicting the ones that became converted to the terrestrial planets; Jupiter more or less followed the general course of the approach of the whole system. The last steps in the capture of primordial Saturn, Uranus, Neptune and perhaps Pluto probably occurred from distances farther away from the Sun than the distance of Jupiter from the Sun. On the other hand, the changing of the orbits of the precursors of the terrestrial planets most probably took place from the side of Jupiter nearest to the primordial Sun. The final rates of orbital velocities of all the protoplanets were determined by their new orbital distances, the closer the orbiting mass to the proto-Sun, the greater the velocity. Although the formation of the orbits of the outer planets occurred with relatively little interference from the satellites of the primordial Sun, the terrestrial ones had to go through many changes in their motions before their orbits became established; the latter group had to pass many times through very high concentrations of antagonistic gases with consequential changes in their compositions and the rates of their motions.

In order to establish the contemporary positions of the planets' orbits on a plane, the equatorial planes of the two primordial systems had to intersect in such a way that the primordial Jupiter's equatorial plane would pass through the central region of the primordial Sun. The angle between the two planes, provided it was not in the neighborhood of zero, does not seem to be of great importance. In the diagram we have arbitrarily placed the equatorial plane of the primordial Jupiter system perpendicular to the plane of the primordial Sun system. However, the important point is: *irrespective of the angle, all members of primordial Jupiter system had to cross the primordial Sun system's equatorial plane twice for each of their revolutions around the primordial Sun; these penetrations through the antagonistic regions resulted in the generation of large amounts of energy which brought about continual reductions in their orbital speeds and the ellipticities of their orbits. After several hundred million years all orbits (with the exception of Pluto's) became almost circular.*

The Second Evolutionary Stage —
The Adjustments

The capture of the primordial Jupiter system produced a very confused picture. We shall attempt to isolate from this chaotic scene some of the major events that led to the emergence of the solar system. However, it should be kept in mind that these events were not isolated cases; many of them were occurring at the same time and affecting each other.

The first significant result of the capture was the entanglement of the small satellites of the two systems, which resulted in a number of +H-antiH reactions and the evaporation of the small frozen masses. In turn, this caused the mixing up of antagonistic gases and the generation of great amounts of heat and low-energy radiation (see conditions for production of radiowaves in the previous chapter). The process then progressed among the somewhat larger frozen masses toward the central region of the newly formed system.

The gases resulting from the evaporation of the small frozen masses were attracted to the larger satellites of the Sun (they dominated the scene). If the atoms in a gas were of the same type of matter as the material of the satellite, they joined it, covered the frozen mass and became its homogeneous atmosphere. If the gas was composed of antagonistic atoms, they formed a shell-like cover above the homogeneous atmosphere (some of the surface mass would have evaporated) and caused a continuous generation of energy and the gradual evaporation of the frozen mass. However, the gases were usually a mixture of the two (from small satellites), in which case the atoms of the same type as the attracting mass penetrated through the tenuous antagonistic layer and joined the homogeneous atmosphere, with the antagonistic gases forming a cover over the structure.

Three principal factors determined the speed with which the frozen masses turned gaseous: their size, the quantity of the antagonistic gases surrounding them and their positions within the system. The last factor determined the amount of the antagonistic masses that could be accumulated. The eventual outcome may be summed up as: all the satellites of the primordial Sun turned gaseous—the smaller ones sooner than the larger ones.

As long as the antagonistic shells surrounding the medium-sized and larger *antimatter* satellites remained intact, their evaporated contents remained contained. But as soon as the $+H$ covers became exhausted, which happened frequently, the contents became released and the gases drifted toward the primordial Sun, attracted by its gravitational forces. Since these gases were composed of *antiH*, they joined other *antiH* gases from the smaller satellites, surrounded the great frozen mass and became its homogeneous atmosphere; the atmosphere became deeper and deeper as more and more of the gases were pulled in. This heavy cover insulated the frozen primordial Sun, while the Sun's mass held the two interacting systems together. As time passed and more and more material from the homogeneous satellites was added to the atmosphere, the central frozen mass became more or less immune to the events in its surroundings. It remained frozen and held the system together until all the newly acquired $+H$ satellites had completed the preliminary stages of their evolutions.

Meanwhile, the reverse of what was happening in the center was taking place in the outer regions of the newly formed system. Here the $+H$ atoms outnumbered the *antiH* ones by a large margin; also, the mode of distribution and the sizes of the larger satellites in the system were different. In the system's central region the *antiH* satellites were massive and congested near the primordial Sun, with the result that the released gases moved toward the central body, joined its atmosphere, and increased its total mass. In the outer regions the larger $+H$ satellites were relatively small and were scattered. Thus most of the released $+H$ gases from the smaller satellites were free to be pulled in toward the center. Consequently, a ring-shaped cloud composed of $+H$ gases was formed along the primordial Jupiter's equatorial plane; the ring was being pulled in by the primordial Sun; as a result it gradually contracted and in the process accumulated more $+H$ gases.

The large primordial satellites from the captured system, including the primordial Jupiter, Saturn, Uranus, Neptune and probably Pluto, had established their orbits far away from the central region and consequently had encountered only small amounts of *antiH* gases. These formed very tenuous antagonistic atmospheres around the frozen masses above their homogeneous atmospheres, produced only minimal amounts of $+H$-$antiH$ *reactions*, and did not generate enough energy to vaporize the contents of these satellites. Therefore, in contrast with the satellites of the primordial Sun system that had turned totally gaseous, the outer planets remained frozen with only small portions of their surface materials turning gaseous. The structure of these planets at that time may be described as a large frozen mass in the center, a shell of a comparatively tenuous $antiH$ atmosphere on the outside and a layer of $+H$ gases in the middle, providing insulation for the frozen central mass. The $+H$-$antiH$ *reactions* between the homogeneous and the antagonistic atmospheres, weak as they were, left their marks by producing atoms of carbon, nitrogen, oxygen, sulfur, etc., which by reacting with hydrogen produced most of the simple molecules now observed on the surface of these planets. We shall describe more fully the effects of the antagonistic atmospheres in the evolution of these outer planets in the next section.

Formation of the Terrestrial Planets

The large primordial satellites of the primordial Jupiter that spiralled toward the primordial Sun (represented by the Earth in the Figure 8 diagram) were crossing the concentrated antagonistic regions of the primordial Sun system continuously. At that time these satellites were large masses ranging between Saturn and Neptune and their gravitational forces were great enough to attract and hold relatively dense and confining $antiH$ atmospheres. Consequently they were heated to incandescence and were transformed to starlets. After the establishment of their orbits, these protoplanets kept orbiting around the primordial Sun, converting part of the contents of their interiors to elements of higher atomic weights (the mechanism described in Chapter IV). During this period they also lost a great part of their masses, as the hot hydrogen atoms could escape through the antagonistic atmosphere. The escape of $+H$ gases became particularly significant in the latter part of the protoplanets' evolution when the radiation zones became weak. During this turbulent period all the medium-sized, and some of the large

$+H$ satellites were lost; the resulting gases escaped into the space near the primordial Sun, and by reacting with the *antiH* clouds from the primordial Sun system's satellites produced enough energy and turbulence to destroy the angular momentum of the *antiH* gases moving toward the Sun. Out of all the satellites of the primordial Jupiter system spiralling toward the Sun only Mercury, Venus, Earth, Moon, Mars, Mars' two small satellites and the "Asteroid's protoplanet" survived the encounter.

The general picture of the evolving system at this time may be described as follows:

1. At the center was a very large, frozen *antiH* mass covered by a very deep homogeneous atmosphere. The frozen mass was rotating inside the atmosphere. The outside portion of the atmosphere was in a turbulent state and had lost its primordial motion. The middle portion between the two had its maximum rate of rotation next to the frozen surface, with the rate gradually diminishing with height above the frozen mass.

2. Clouds of gases composed primarily of *antiH* were drifting toward the center of the system. Interdispersed among these clouds were small clouds composed principally of $+H$ gases; these were also moving toward the center, with the radiation between the antagonistic clouds keeping them apart. The *antiH* clouds were advancing along the primordial Sun's equatorial plane; almost all of the $+H$ clouds were situated on the primordial Jupiter's equatorial plane.

3. The frozen outer planets were orbiting around the Sun along Jupiter's equatorial plane; each of them had a tenuous *antiH* cover over its homogeneous atmosphere, the antagonistic material having been collected from the primordial Sun's small satellites after their conversion to the gaseous state.

4. The inner protoplanets were also orbiting around the proto-Sun along the Jupiter's equatorial plane. These newly acquired satellites had to pass through the *antiH* clouds in the primordial Sun's equatorial plane—twice for each revolution around the proto-Sun. Each time they passed through the clouds they transferred some of their angular momentum to the gases which in turn caused the slowing down of their

orbital velocities. The protoplanets were transferring the kinetic energies they had acquired from the primordial Sun, during the capture, to the *antiH* clouds. The total effect was that these gases were being gradually pulled into the Jupiter's equatorial plane and given motions in the same direction as the motions of the protoplanets.

5. Another cloud of gases, composed of $+H$ gases, had gathered outside the orbit of Neptune in the form of a ring coplanar with Jupiter's equatorial plane. This cloud, being the product of the evaporation of the primordial Jupiter's small satellites, was rotating in the same direction as Jupiter's direction of rotation and carrying with it a considerable amount of angular momentum. This ring was contracting slowly, pulled in by the gravitational forces from the center, and accumulating the $+H$ atoms in its path.

To sum up, the capture of the primordial Jupiter system had given large quantities of kinetic energy to the inner protoplanets; this energy, together with the annihilation of part of their masses, was utilized to give the *antiH* gases a new direction of rotation along Jupiter's equatorial plane. This change of direction, together with the tidal effects caused by the orbital motions of the outer satellites, was gradually turning the directions of motions of the gases in the proto-Sun's upper atmosphere (they had already lost their initial angular momentum) toward Jupiter's equatorial plane.

The *antiH* clouds originating from satellites continued to be pulled in by the central mass, followed by the contracting $+H$ gas ring that originated from Jupiter's small satellites. These gases were spread out through vast regions and, similar to the nebulae in our Galaxy, tenuous. The two major antagonistic gaseous clouds were kept separate through the effects of the radiation zone at their boundary; the continuous generation of energy in this region caused the outer ring to expand, cover up the *antiH* cloud from all directions, and form a shell over the whole structure. With the progress of contraction, the *antiH* gases slowly receded and their outer boundary passed the orbits of the inner planets one by one. This meant the protoplanets could no longer replenish the used-up *antiH* gases in their antagonistic atmospheres. However, the gravitational forces of the inner protoplanets had retained considerable amounts of *antiH* gases in their upper atmospheres; it took a fairly long time before these antagonistic atmospheres became totally exhausted.

During this period the gradual reduction of the concentration of *antiH* atoms caused the temperatures of the terrestrial planets to drop slowly; also, during this period most of the free hydrogen and helium atoms and molecules escaped through the weakend radiation zone. By this time the physical states of the planets had altered considerably. The most important changes were: (1) most of the original mass had been lost; (2) a considerable amount of hydrogen had become converted to heavier elements; and (3) large quantities of water vapors had been produced by reactions between hydrogen and newly formed oxygen atoms—the water molecules were too heavy to pass through the radiation zone and remained on the planets until they cooled.

Formation of the Proto-Sun

The cooling down of the planets more or less coincided with the gradual warming up of the proto-Sun. But this was not a mere coincidence; the same conditions which caused the cooling down of the planets were helping to heat up the central mass. Most of the *antiH* clouds had joined the homogeneous atmosphere of the Sun, and the rest were in the process of being assimilated. The whole unit was then surrounded by the $+H$ gases that had followed the contracting *antiH* clouds. The diameter was becoming smaller and smaller and the radiation from the zone between the two antagonistic gases was becoming stronger and more effective. After a long time, all the frozen mass in the center gradually became gaseous, and the energy from the radiation zone heated the central body to incandescence. Finally, an equilibrium was established between the interior gases and their antagonistic atmosphere where the rate of annihilation of atoms was controlled by the amount of energy produced within the radiation zone. We shall elaborate on this aspect of the equilibrium presently.

After their evaporation, the *antiH* gases in the interior of the central mass continued to rotate along the original primordial Sun's equatorial plane, but these gases could not retain the primordial Sun's angular momentum. A great amount of the primordial angular momentum had already been dissipated; then the continuous changes of velocities and directions of movements of the atoms as they bounced against the radiation wall (brought about by the annihilation of atoms between the two antagonistic layers) destroyed what was left of the original angular momentum. However, the gases on the upper atmosphere of the Sun were subject to the tidal effects of the orbiting planets, especially that of the massive Jupiter. This dragging process made the surface gases along the circle located at the intersection between the surface of the Sun and the equatorial plane of Jupiter develop a rotational motion which spread out toward the poles. This type of motion is still in effect; the present equator of the Sun, situated on the common plane shared by the orbits of the planets, is rotating at a faster rate than the rest of the sphere. Therefore, the present equatorial plane of the Sun is in reality the same as the equatorial plane of the primordial Jupiter and as far as the rotational motion and angular momentum is concerned, the Sun is subservient to its planets. The changing of the position of the equatorial plane of the Sun had another effect: the axes of rotation of the Sun and Jupiter became almost parallel.

A significant conclusion with practical implications may be drawn from these discussions: *the probabilities are that any star bearing large planets with orbits on a plane intersecting the equator of the star has a nonuniform pattern of rotation similar to that of the Sun.*

The Third Evolutionary Stage — Progress Toward Stability

On examining the curve in Figure 4, depicting the general pattern of evolutionary trends, one finds no abrupt transformation from the second to the third stage; instead, the two stages phase into each other. Also, the third stage is a very slow and prolonged process. As compared to the stars of the nucleus of the Galaxy, the solar system is relatively young and may be considered as an example of a typical star that has gone less than a quarter of the way toward reaching its best possible state of stability. In the following discussions we shall consider the solar system, and its properties, as an average stable star that is progressing along its third evolutionary stage.

Before we embark on the discussion of Sun's properties, let us take a look at what happens when $+matter$ atoms collide with *antimatter* atoms. It is evident that the annihilation does not result immediately in the generation of equivalent amounts of electromagnetic energies. Electromagnetic waves are produced by the oscillation of electrons in atoms and therefore cannot be the direct product of such reactions. The immediate product is most probably a mixture of *force-* and *non-wave-energy-* primatons. These are the reasons: the formation of a pair of symmetrical particles requires the introduction of appropriate amounts of energy in a field of force—that is, pair production is the outcome of interactions between forces and energies. Thus when two symmetrcial particles annihilate each other (the reverse of the process of pair formation), both *energy- and force-* primatons should be produced.

If this is the case, then the gravitational attraction of a star is not due entirely to the gravitational force-primatons, but to a mixture of weak and strong forces—the greater the rate of energy production the higher the percentage of the strong forces. Therefore, not only does the annihilation of matter continually cause gradual increases in gravitational attractions (Chapter IV), the intensity of energy production is a measure of additional and stronger force-primatons emitted at any given time. These variations in the values of gravitational constants of stars cannot be recognized by our present methods for the determination of the strengths of gravitational attractions of stars. The usual procedure is to calculate the gravitational forces from the motions of stars, accept a universal value for gravitational constant, and attribute the calculated results to the masses of the stars. This is a sound practice for practical purposes, but does not provide us with the true picture of evolutionary trends. As will be seen in Chapter X, the realization of the presence of strong force-primatons among gravitational forces is an important step in following the trend of evolution of matter.

With regard to the produced non-wave energy primatons, most of them are used soon after their appearance in the excitation of atoms and are converted to electromagnetic energies. However, a great possibility exists for some of the non-wave energy-primatons to pass through without being transformed to waves. In this respect it would be interesting to measure above the Earth's atmosphere all the energies received in the form of electromagnetic waves from the Sun on a specified area and then compared with the total amount of energies received in the same area. In such experiments we should first define what is meant by wave and non-wave energies. For example, should gamma-rays be placed under the non-wave category or accepted as true electromagnetic waves?

Sun

The Sun consists of a large sphere of incandescent *antimatter* gases surrounded by a tenuous +*matter* atmosphere composed primarily of hydrogen. In the region between the two antagonistic gases annihilation of matter is taking place, with the produced energy counterbalancing the gravitational effects of the central body on the atmospheric gases and keeping the equilibrium intact. The radiation zone between the two gases controls the rate of penetration of the atoms of one type into its antagonistic region. The strength of radiation is not uniform over the sphere, but regional. In a specific location the penetration of the +H atoms occurs when the gravitational forces are more powerful than the radiation; the penetration of the *antimatter* takes place when such surface activities as flares and prominences carry the internal material with them into the atmosphere. When excessive amounts of annihilation occur in an area, large amounts of energy are produced in the region and the radiation pressure causes a reduction in the passage of atoms through the radiation zone. On the other hand, when in a region insufficient amounts of annihilation are taking place, the drop in the radiation pressure allows more atoms to pass through the zone. Consequently, the overall rate of annihilation is very steady and is controlled by the total mass of the Sun. The mass determines the concentration of the atoms in its atmosphere through its gravitational attraction, which in turn is counterbalanced by the repulsive effects of the produced radiation. The interactions and counteractions have created a delicate balance between the main body of the Sun and its atmosphere.

The distribution of gases in the interior of the Sun is not uniform. The concentration of atoms is greater in the center than near its surface; the reason for this state is that in a confined sphere there are statistically more atoms in the center than near the barrier causing the confinement. But this does not mean that the center of the Sun is under such a tremendous amount of pressure to bring about the fusion of atoms.

If we accept the concept that the heavy elements of the terrestrial planets are the products of +*matter-antimatter reactions*, then we should also expect the interior of the Sun to contain a very high percentage of *antimatter* heavy elements. In evaluating this possibility the following three facts should be considered: (1) the Sun did not lose its *antiH* gases as the terrestrial planets lost their +*H* gases; (2) through its +*matter-antimatter reactions* the Sun has been converting its contents to heavier elements for a much longer time than the terrestrial planets and consequently most of the original *antiH* atoms have been transformed into heavy elements; and (3) the gravitational constant of the Sun has been increasing for the past five and a half billion years.

Since this higher value is due to an increase in the mass of the subatomic particles (Chapter IV), all the atoms in the interior of the Sun are most probably denser than their counter-equivalents on the planets. Therefore, a great possibility exists for the gases of the interior of the Sun to be composed primarily of dense and heavy atomic nuclei which are holding together through the attractive forces between the atoms themselves; the radiation zone merely adds stability to the unit. The energy from this region on one hand keeps the contents gaseous and on the other helps to hold the unit intact. This was not the case in the early stages of the Sun's evolution. The principal function of the radiation zone at that time was to confine the light gases; then the Sun had a larger diameter than at the present.

The photosphere, the region on the surface of the Sun from where radiation is emitted into the outer surroundings, is opaque and thus generates a typical black-body radiation of continuous spectrum. In the prevalent theories the cause of opacity is attributed to the occasional presence of negative hydrogen ions (hydrogen atoms possessing two electrons). Accordingly, it is assumed that while the hydrogen gases extend into the corona and beyond, the formation of the negative hydrogen ions stops at the outer regions of photosphere. Though this stoppage can be explained theoretically, no convincing reasons have been given as to why the eruptions on the surface do not cause the penetration of the hot, negative ions into the regions above, and give the Sun a diffuse edge. The phenomenon of "limb darkening" is then regarded as a confirming evidence for the negative hydrogen ion theory and used for the theoretical construction of the structure of the interior of the Sun.

Here we are suggesting an alternative explanation in which the energy of the Sun is generated on its surface, the interior of the Sun is composed of heavy and dense atoms, the concentration of atoms per unit volume increases toward the center and finally, the opacity of the main body of the Sun is caused by the heavy and dense atoms. The darkening of the limb is then explained by the fact that the higher concentration of atoms in the central part of the sphere produces a more intense radiation in that region, and the geometry of the sphere relative to the observer causes the reception of smaller amounts of light from the limb.

Solar Atmosphere

In contrast to the dense composition of its interior, the atmosphere of the Sun is tenuous. Yet the atmosphere is not a continuum throughout; it is an intricate gaseous structure with many parts, each having its own specific properties. The outermost atmosphere may be divided into three main layers, the chromosphere, the inner corona, *K corona*, and the outer corona, *F corona*. The chromosphere has a recognizable and distinct boundary with the photosphere, but gradually phases into the inner corona, which in turn imperceptibly extends into the outer corona. In spite of this gradual transformation, each of these atmospheric parts has a characteristic of its own.

We have evidence showing the extension of the Sun's atmosphere to beyond the Earth's orbit, but there are no reasons to assume that it ends there. The outer corona gradually becomes more tenuous with distance, but inside this continuum there are currents of subatomic particles that are intermittently ejected by the Sun (solar wind). The atmospheric gases, whether near or far, are all under the influence of two principal factors, the attractive gravitational forces of the Sun and the repelling radiation from its surface. The balance between the two has caused the concentration of the atoms in the corona to be relatively low and to become more tenuous as it extends into space. No limits can be assigned to the extent of this ever-thinning atmosphere, but the possibilities are that it extends far beyond the orbit of Pluto. The material in the solar wind cannot escape the gravitational pull of the Sun, and after the particle energies are expended they become the unidentifiable part of the system. Now the equilibrium of the Sun requires that the annihilated $+matter$ gases be replenished continuously; this means that there is a continuous and imperceptible movement of the atoms from the outer regions toward the Sun. The Sun still has a vast amount of $+matter$ supply. In estimating the quantity of the available material and hence the radiating life of the Sun, one should keep in mind that the consumption of its atmospheric gases is quite thrifty and very efficient. The produced energy comes from total annihilation of matter and only half of this matter is from the atmosphere.

The material of the corona moves in two main directions: first, outward in streams (the continuation of the effects of reactions within the spicules) in which the electrons (the major component), protons and atomic nuclei are driven toward the outer regions by the intense radiation energies; and second, inward, a general motion of the atoms toward the central mass in between the streams, the movement being caused by the great gravitational forces of the Sun. The gases pulled toward the interior are composed of $+H$ mixed with relatively small numbers of higher atomic weight elements (the origin of these will be explained shortly). These gases move in between the outward streams toward the low radiation pressure areas on the photosphere. The nearer they get to the photosphere the greater are their velocities, and by the time they reach the chormosphere the nuclei (the electrons have been knocked off by radiation) have sufficient momentums to pass through the relatively weak radiation zone and reach the photosphere.

The intensity of the resulting eruption depends on the size of the weak pressure area, the extent of the weakness in radiation pressure and the amount of surging +*matter*. These are the possibilities for different types of eruption: (1) in most cases the onrushing gases pass through small weak areas and, by reacting with *antimatter* material, produce X-ray emitting granules on the photosphere; (2) if the +*matter* gases enter medium-sized areas they cause the formation of flares; and (3) if the receiving area is large, great quantities of +*matter* surge in, and a huge explosion in the form of a prominence results. The X-ray emitting eruptions may be regarded as reactions on the surface of the photosphere with the produced radiation pressure sufficient enough to prevent further entry of the +*matter* gases for a short time. The flares are more violent and cause the reversal of the motions of part of the onrushing +*matter* material; these medium-sized eruptions produce outward streams of what is left of the +*matter* particles in which some of the *antimatter* mass is also carried; the interactions that occur between the antagonistic atoms within the produced outward streams inside the chromosphere produce the larger spicules.

The large eruptions which produce the prominences cause an outward thrust of a considerable amount of the Sun's material. But as the *antimatter* material penetrates into the antagonistic coronal region, it encounters a wall of radiation created by the intensified +*matter-antimatter reaction*. The emission lines of ionized helium are generated in this region. The outward rush is eventually brought to a stop by the combination of the effects of the radiation pressure in front and the pull of the gravitational forces from behind. The prominence then virtually splits into two dissimilar parts. The *antimatter* material, being dense and not encountering much resistance on its return trip, pours back into the Sun; the +*matter* subatomic particles are ejected outward, pass through the corona and produce powerful solar winds.

Explanations of Some of the Observations

Let us now attempt to correlate the proposed hypothesis with some of the observations. The most important of these are the spectrographic data related to the Sun and the various parts of its atmosphere.

To begin with, the continuous spectrum from the photosphere may be ascribed to the radiation from the hot, heavy and dense atoms inside the Sun. The X-ray-emitting eruptions among the granules on the surface of the photosphere may be attributed to small-scale annihilations on the surface; in these regions the *+matter* particles (mostly atomic nuclei) enter the areas in scattered bunches, with each particle producing a speck of intense radiation at the point of collision.

In addition to the lines of hydrogen and other heavy nuclei, we recieve the emission lines of helium from the chromosphere, indicating the presence of sources of very powerful energies in that region. Yet we know the chromosphere to be highly tenuous; this infers the production of intense energies by a relatively small number of atoms, the type of energies that are produced by total annihilation. The fact that the temperatures of the upper chromosphere are within the neighborhood of 100,000° K while the densities of the gases decrease upward can be taken as an indication that most of the annihilation within the chromosphere is taking place in the lower half of the chromosphere in between the photosphere and the inner corona.

A flaw seems to be present in this hypothesis: it implies that no heavy atoms should be present in the chromosphere. This is the reasoning for the implication: the gravitational attraction of the Sun is much more effective on the heavier nuclei than on protons and electrons, and hence during the long life of the Sun, these atoms should have been preferentially pulled into the photosphere and annihilated. But we know this not to be the case and thus have to provide an explanation for their continued presence. The theory mentions that within the spicules a mixture of gases composed mostly of protons and some *antimatter* material from the Sun is surging outward, driven by the energies emanated from the base areas on the photosphere. As the mixture moves within the streams, antagonistic materials interact and the resulting energy causes the conversion of some of protons into heavier elements. The principle of this transformation is the same as the one described for the formation of heavier elements in the terrestrial planets. The energies produced by the +*matter-anti-matter reactions* reinforce the radiation from the photospheric region and accelerate the motions of the newly formed nuclei, protons and electrons toward the corona. Their motions slow down somewhat in the outer corona but their momentums carry them through into space. Gradually they become scattered and are pulled back. But these nuclei, being more massive than the rest are selectively attracted, join the gases that are destined to move toward the Sun and eventually become annihilated. Therefore, the seemingly permanent existence of the heavier elements in the chromosphere is the outcome of the continual conversion of the hydrogen atoms to the heavier units, which after a long journey into space return to the Sun and become destroyed. Here a balance has developed between the rates of creation of the heavier elements and their annihilation, and as a result, their relative abundance remains about the same all the time.

No distinct boundary can be observed between the chromosphere and the corona, for the former merges indistinctly into the latter. However, this gradual transformation is accompanied by considerable changes in the properties of the gases and the particles. Whereas the atoms in the chromosphere generate emission lines, the gases in the inner corona emit a continuous spectrum; even the familiar lines such as the H and K lines of calcium are absent. In the outer corona the spectrum changes again. From this region we receive not only broadened and distorted lines but also the lines of highly ionized atoms of iron, nickel and calcium, indicating the presence of extremely energetic activities.

The sunlight is scattered in the corona and is partially polarized; these effects are probably brought about by the motions of high-speed subatomic particles, principally electrons. The presence of the emission lines of ionized helium and other elements in the upper chromosphere, the absence of any lines from the inner corona (indication of very fast-moving electrons), the emission of highly ionized iron, nickel and calcium lines in the outer corona (as will be seen shortly, evidence of slowing down of

the particles), the "pearly appearance" of the corona (implying the presence of large number of streams of particles) and
the ejection of particles into space in the form of solar wind
all indicate a continuous outflow of electrons. Their point of
origin seems to be in the upper chromosphere; the electrons apparently reach their maximum velocities in the inner corona and
slow somewhat in the outer corona. The question is: what is their
source?

Mention has been made of the presence of two main types of
motions in the corona: streams of particles (mostly electrons)
moving outward and scattered atoms (mostly hydrogen) moving
toward the Sun in between the streams. Because the atoms
from the latter group have to pass through regions of powerful
radiation, their electrons become stripped from them and are
propelled outward. Being much heavier, most of the protons and
nuclei continue with their motions, cross the areas of low-radia-
tion pressure and either become annihilated on the photosphere
or turned back by the radiation. The electrons, being very light,
generally do not follow this course; the radiation produced by
the annihilation of their nuclei drive them (and some of the late-
arriving protons and nuclei) toward the corona. With the powerful radiation behind them, the electrons gain speed and reach
their maximum velocities in the inner corona, velocities that
cause the obliteration of all the spectrographic lines. As the electrons move outward and their distances from the sources of radiation increase, their velocities become reduced. It is believed that
this reduction in speed is responsible for the type of spectrum received from the outer corona.

The heavier nuclei, such as iron, nickel and calcium, created in
the chromosphere and driven outward by radiation, move inside
the streams of high velocity electrons and consequently do not
have the ability to emit their specific spectral lines in the inner
corona. However, by the time these nuclei reach the outer corona
they are capable of repossessing about half of the normal number
of their electrons; the number they can recapture depends on the
balance between the electromagnetic forces of the nuclei and the
velocities of the electrons. We know the closer the orbit of the
electron to the nucleus the stronger the attractive forces; therefore, the observed spectral lines of the highly ionized atoms are

not in reality a measurement of the temperature of the region, but a measure of the velocity of electrons in relation to the strengths of the forces present in the different orbits in these elements.

We have not included the sunspots in this discussion as they do not seem to be related directly to the evolution of the Sun. The sunspots are not permanent features and thus cannot be accepted as important components of the mechanisms which produce solar energy. Their importance as indicators of the solar activities cannot be denied. However, their inconsistent periodic occurrences, their cyclic changes of polarities of their magnetic fields and their concentration around the equator, where the gases move at a faster rate than the rest of the surface gases, relegates them to phenomena resulting from disturbances within the equilibrium of the Sun. Since they do not contribute continually toward the maintenance of the equilibrium, they cannot be considered of great evolutionary importance.

By now it is evident that the containment of the gases in the interior of the Sun is due to a number of factors: the heavy and dense atoms, the very strong solar gravitational forces, the enveloping radiation pressure surrounding the sphere and the antagonistic effects of the coronal material which prevents the *antimatter* material from penetrating too far into the region. These are formidable barriers which make it highly unlikely for some of the *antimatter* atoms to be ejected into space. Nonetheless, there is always a possibility of an escape, especially in conjunction with the forceful thrusts of the very large prominences. Therefore, consideration should be given to the probability that some of the captured *antimatter* nuclei discovered in the cosmic-ray experiments are coming from the interior of the Sun.

In concluding this section, mention should be made of the expected life of the Sun. It is obvious that the solar atmosphere is the least durable component of the equilibrium and hence the first to become exhausted. However, we cannot make any estimate as to the length of time the atmosphere will last. Since we do not know from how far in space the Sun can pull in the $+H$

atoms, we are ignorant of the quantity of the reserves; neither do we have any knowledge of what roles the frozen satellites of the outer planets are playing in the replenishment process. Above all, we cannot predict the influence Man may have in converting the frozen planets to solar atmosphere (Chapter XI). However, it is certain that some time in the distant future all the supply of +*matter* gases will be consumed and the production of energy will come to a halt. Naturally, the reduction in the concentration of the atmospheric material will be at a slow rate and the cooling will be very gradual. This is the fourth stage of the evolution of the Sun and very similar to the expected fate of most of the red dwarfs (details in Chapter VI). At that time our most precious source of energy will become transformed to a very dense, spherical, solid mass. Then after an extremely long interval, it may get its chance to contribute its trifling share toward the dissolution of the Galaxy.

Terrestrial Planets

Since all the inner planets are presumed to have passed through the same evolutionary stages, we may attribute the major differences in their present properties to the occurrence of isolated incidences during their capture. Their rates of rotation provide the best example. It may be said that the primordial Mercury and Venus lost their original rotations by having come in contact with *antimatter* satellites. This supposition is strengthened by the following possibility: the nearer the protoplanet was to the center of the system, the greater its chance of encountering one of the primordial satellites of the Sun. However, the likelihood of head-on collisions was extremely small. In most cases the contacts were of a glancing type, as energies produced in such encounters would have caused the separation of the two bodies without producing gigantic explosions. Nevertheless, these surface +*matter-antimatter reactions* had enough power to bring about the loss of the rotational motions—and if there was more than one encounter, a change in the direction of rotation (this may account for the probable cause of the retrograde rotation of Venus). The contemporary mode of rotation of Mercury and Moon may then be ascribed to the effects of the physical features developed later on during their evolution (e.g., asymmetrical distribution of dense material within the sphere).

It has been stated that the last stages in the establishment of the equilibrium of the Sun corresponded with a gradual recession of the *antiH* clouds toward the center of the system. This means that the *antiH* atmosphere of the outer terrestrial protoplanet was the first to become exhausted. Thus the order of exhaustion (if we assume the protoplanets were approximately of the same mass) would have been Mars, Earth, Venus and Mercury. However, the present masses of the planets indicate variations in their primordial masses, with the primordial Earth having been the largest of the four and the primordial Venus the second largest. The larger the primordial satellite the greater the amount of *antiH* clouds it could retain; therefore, Earth and Venus most probably held on to their antagonistic atmospheres longer than Mars and our Moon. Since Mercury was closest to the Sun, irrespective of its mass, it was the last to lose its antagonistic atmosphere and hence was in contact with *antimatter* just as long or even longer than Venus or Earth. The crucial point in this discussion is that the longer the protoplanet remained in contact with its antagonistic atmosphere, the greater was the possibility of its lighter elements becoming converted to the heavier elements.

The atmospheres of the planets have contributed a great deal toward the formation of the contemporary surface features on the planets. In return, most of the properties of the atmospheres, such as the surface pressures and chemical compositions, are in great part the outcome of the evolution of the planets and hence important clues to their past. Of the five large terrestrial satellites, two, Mercury and Moon, are without atmospheres. Mars has a semblance of one, Earth's atmosphere has completely altered from its original form, and Venus is surrounded by a heavy blanket of gases composed primarily of carbon dioxide. The accounting of the causes of this diversity is a major step toward understanding the path of evolution of the terrestrial planets.

The basic assumption in these hypotheses has been that all the inner satellites of the Sun have gone through the same processes of elemental transformations, and that by the time they had begun to cool down they all possessed atmospheres containing the same types of compounds. Most probably the atmospheres were composed of (though the percentage relationship was not necessarily the same) hydrogen, helium, water vapors, carbon dioxide, carbon monoxide, methane, ammonia, hydrogen cyanide, hydrogen sulfide and other simple inorganic and organic compounds. In the satellites' molten stage, the radiation zone between the homogeneous and the depleting antagonistic atmospheres contained the gases and prevented their escape. In the latter part of the cooling period when the radiation zone became weak, most of the free hydrogen and helium were lost, but the rest of the gases remained with the satellites until the radiation zone disappeared entirely, which occurred after their surfaces had become solidified. Therefore, the observed variations in the states of atmospheres of Mars, Earth and Venus are related to the post-cooling era.

We do not have to look far to find the reasons for the total disappearance of the atmospheres of Mercury and the Moon. It is obvious that the prime cause was their small masses. The two small satellites could not retain the surface gases after the confining effects of their radiation zones gradually became ineffective. The lack of evidence for the presence of bodies of water during the geological times on either the Moon or Mercury implies the loss of the atmospheric gases at relatively fast rates. This is substantiated by the fact that both satellites rotate at very slow rates, permitting excessive collection of heat in certain regions for long periods of time; most probably this has caused the escape of the gases at fast rates from the hot side of the satellite.

The planet Mars is about twice as massive as Mercury, but this difference alone does not explain the cause of the slow depletion of its atmosphere. We can be quite certain that in the beginning the planet had an atmosphere comparable to that of Earth. The reasons for this statement are the discovery of the dried-up channels with tributaries on the planet and the fact that Mars still possesses a tenuous atmosphere. The valleys and channels show the great probability of the presence of relatively large amounts of water on the young planet, which in turn suggests that most probably the atmosphere contained a variety of gases such as methane, ammonia, hydrogen cyanide, carbon dioxide, etc. The cause of the slow loss of the atmospheric gases may be attributed to three principal factors: a large enough mass to prevent the fast escape of gases; the reception of reduced amounts of radiation from the Sun due to distance; and the relatively fast rate of rotation of the planet, which has caused a fairly even distribution of energy over the surface of the planet.

As far as the atmosphere is concerned, the Earth was at the most advantageous position right from the beginning. Its mass was large enough to hold on to most of the gases and water vapors on its surface after its antagonistic atmosphere became exhausted. It was at a proper distance from the Sun to receive adequate but not excessive amounts of energy. And it was rotating at the proper rate to insure mild climatic conditions. These factors brought about a stability which was one of the requirements for the origin of life. The evolution of life in turn changed the composition of the atmosphere and established a delicate system for the utilization and disposal of the energy received from the Sun. We shall come back to this subject in the last chapter, where the role of the atmosphere in the origin and evolution of life will be discussed in more detail.

In many respects the young Venus was similar to the cooling Earth, but it differed in two major categories: it was closer to the Sun and thus received a more intense radiation; it was rotating at a very slow rate, and became subject to excessive heating of the side facing the Sun. Most probably these two differences were the factors which set the evolutionary trend of Venus toward becoming an inhospitable hot planet enveloped in a dense atmosphere composed of carbon dioxide, carbon monoxide and small quantities of water and corrosive acids.

If we assume that all the terrestrial satellites began with atmospheres containing approximately the same chemical contents, then the original atmosphere of Venus would have contained large quantities of water. But since the planet was rotating very slowly, any point on its surface was being subject to continuous and intense radiation for 243 Earth days. The section coming into view of the Sun became hotter and hotter with time and after eight months of relentless radiation lost some of the lighter compounds (water and compounds of similar molecular weight) from the atmosphere to space. But the continuous and intense radiation had other deleterious effects: it also caused the dissociation of compounds to their elemental components which could escape at faster rates than the undissociated compounds. Four and one-half billion years of this type of depletion can result in great changes in the relative abundance of the compounds. The gases that were not particularly affected by these conditions were carbon dioxide and carbon monoxide. Both being stable, heavy compounds, these gases became the major components of the atmosphere. As their relative concentration increased, the planet came more and more under their greenhouse influence. Eventually the temperatures reached the levels at which the breakdown of the carbonate rocks occur. The gradual release of carbon dioxide gases then resulted in the dense atmosphere of the contemporary Venus.

Before completing the discussion on the terrestrial planets, mention should be made of the possibility of a direct relationship between the cratering phenomenon and the last stages of the evolution of the Sun. Most of the evidence about cratering comes from the Moon, with Mercury and Mars providing corroborative information. It is generally agreed that most of the cratering occurred early in the lives of the planets, approximately four billion years ago. But the significant point about this phenomenon is that although isolated craters are more or less scattered uniformly over the surface of the Moon, the mare basins and similar features on Mercury (the Caloris Basin) and Mars (the Halas, Argyre and Isidis Basins in the southern hemisphere) are located only on one side (over about half of the sphere). This pattern of distribution points to the great possibility of the occurrence of a series of impacts about four billion years ago when the largest and most violent ones came from one specific direction. In our present state of knowledge it would be premature to draw definite conclusions as to the source and the nature of the impacting objects. However, consideration should be given to the possibility that in the last stages of the Sun's contraction every so often it ejected some of its contents into space and these *antimatter* objects were responsible for the cratering of the surfaces and the formation of the maria basins. The surface explosions on the Sun that caused the ejections of the *antimatter* materials could have been brought about by impacts with *+matter* planetesimals. The individual *antimatter* meteorites would have then been responsible for the formation of the craters and the large meteor showers would have produced the marias. Most probably there were a number of ejections at irregular intervals with the positions of the satellites with respect to the sites of explosion determining the outcome. If a planet was in the direct path of the thrust (an infrequent occurrence) it was bombarded with large quantities of meteors and meteorites on the side that was facing the source of explosion (maria formation); and if not, isolated meteorites that had scattered around would collide with the satellite at any random location on its surface (crater formation).

Jovian Planets

The prime cause of the differences between the properties of the outer and the inner planets may be ascribed to the amounts of *antiH* gases encountered by the primordial frozen satellites at the time of their capture. Because most of the *antiH* frozen masses were situated within the present orbits of the terrestrial planets, the inner protoplanets were subjected to far greater degrees of intense reactions than the outer ones. But since the amounts of *antiH* gases inside the orbital regions of the Jovian planets were relatively small, no spectacular alterations occurred from the ensuing reactions. As to be expected, the greatest effect was registered on Jupiter and its satellites, the nearest of the Jovian group to the Sun and the ones that were likely to encounter the greatest amounts of *antiH* gases.

Similar to the terrestrial protoplanets, all the captured Jovian primordial frozen satellites had to pass through myriads of *antiH* frozen masses belonging to the primordial Sun system, twice for each revolution around the Sun. But, whereas the concentration of *antiH* gases was high within the orbits of the terrestrial protoplanets and remained high for a long time, the outer protoplanets became covered with relatively tenuous *antiH* atmospheres, with the tenuity increasing roughly in proportion to the distances from the Sun. Not only the sizes and the numbers of the frozen *antiH* masses became smaller with distance, but the resulting gases had to fill up a larger space; hence they were less likely to accumulate around the Jovian protoplanets. The effects of the *antiH* atmospheres may be seen in the chemical composition of the atmospheres of the outer planets—and more importantly—on the densities of the five inner moons of Jupiter.

If we accept the concept that the formation of the heavier elements was the product of $+H$-$antiH$ *reactins* then such elements as carbon, nitrogen, oxygen, etc. (observed in the atmospheres of the Jovian planets in the form of simple compounds) are also products of $+$ *matter-antimatter reactions* that have occurred on the surfaces of those planets. The probabilities are that smaller quantities of still heavier elements were also produced, but because of the higher freezing points of their compounds, they are at present frozen on the surfaces of the planets and cannot be observed in their atmospheres. The significance of the simple compounds found in the atmospheres of the Jovian planets lies in their indication of the presence of antagonistic atmospheres over their homogeneous ones some time ago. There were not enough *antiH* gases to bring about important changes, but sufficient to convert some of the surface hydrogen to the heavier elements.

Now, if the Jovian protoplanets became covered with antagonistic atmospheres, then their secondary satellites should have been continually moving inside antagonistic gases for long times; thus their development more or less should have duplicated the pattern of the evolution of the terrestrial planets. We can see this phenomenon plainly among the Jupiter's inner moons; in fact these satellites can provide us with a fairly accurate estimate of the height of Jupiter's antagonistic atmosphere. We know that the densities of these moons fall off approximately in proportion to their distances from Jupiter, with Callisto, the farthest of the group, having the lowest density. This means that the concentration of *antiH* gases above Jupiter was at its highest between the planet's homogeneous atmosphere and the orbit of Io, becoming more and more rarified with distance, but still effective at Callisto's orbit. Then, as the antagonistic atmosphere was used up, Callisto was the first to come out from under its effects and with the least amount of conversion of its elements to heavier ones. With the reduction in the depth of the antagonistic atmosphere, one by one the moons emerged—and the later the emergence the higher the density.

The following observed properties of the Jovian system suggest the possibility of +*matter-antimatter reactions* occurring at present in the upper atmospheric regions of Jupiter: (1) the higher-than-average surface temperatures of the belts; (2) the presence of colored compounds of unknown composition in the belts; (3) the radiation of twice as much thermal energy from the planet than the amount received from the Sun; (4) the intermittent emission of decametric radiation, the radiowaves originating from untraceable sources; (5) the bursts of radio noises when Io reaches certain positions relative to Jupiter; (6) the presence of atmospheric gases on Io (about the size of our Moon) in the forms of neutral hydrogen around the mass and ionized particles at higher levels; (7) the emission of D-lines of sodium by Io, a satellite which is supposed to have a temperature of about -150° C; and (8) the inconsistencies observed in the ionosphere and the magnetosphere of the planet, with the motions of the inner moons influencing the strengths of the charged particles. These unexpected findings may eventually be traced to causes other than +*matter-antimatter reactions*, but until then the possibility of the existence of such activities cannot be ignored.

If the +*matter-antimatter reactions* are responsible for the as yet unexplainable observations, they are of the type that is occurring within the electromagnetic wave emitting nebulae of the Galaxy. The details of this type of reactions are given in Chapter VII, but in brief it may be stated that the encounter between free moving +*matter* and *antimatter* particles usually does not result in total annihilation. In most cases, the energy produced during the encounter not only brings about the separation of the two, but also causes the surrounding particles to accelerate. On collision with atoms of their own type, the accelerated particles produce a number of smaller charged subatomic particles in the manner the secondary cosmic rays are released in the Earth's atmosphere. These particles then produce some of the observed phenomena.

The probability of +*matter-antimatter reactions* occurring in the upper atmosphere of Jupiter raises the question of the source of *antimatter* particles. Two possible sources come to mind: the first but the least likely one is that some of the primordial *anti-*

matter atoms have been trapped inside pockets within the Jovian atmosphere and are continuously leaking out; the second and the more probable one is that at times *antimatter* particles become trapped inside the upper atmospheric gases, the antagonistic particles coming from the unobservable comets.

The nature of the comets will be discussed in detail in the next section. It will be shown that they are *antimatter* meteorites which shed particles from their surfaces as they collide with +*matter* atoms in the course of their movements around the Sun. But the glows of comets are very weak in the vicinity of the orbit of Jupiter, and they cannot be usually observed. Consequently, unless someone is looking at the Jupiter at the time of the impact, a rather rare event, the collisions of the comets go unnoticed.

An *antimatter* meteorite may collide with Jupiter either by spiraling inward or by direct collision. In the former case, as the comet comes in contact with the atmospheric gases, it sheds particles at a very fast rate until all the mass is used up; in the latter case it leaves its mark at the area of impact in the form of a red spot on the surface of the planet. It should be remembered that such encounters occur only at infrequent intervals. However, once the *antimatter* particles shed from the comets become intermingled with the tenuous +*matter* gases, they remain within the atmosphere for some time. During this period the mixtures are dragged latitudinally by the Coriolis forces of Jupiter (caused by the fast rate of rotation of the planet) and form the brownish-red belts.

The *antimatter* meteorites that collide directly, penetrate through the Jovian atmosphere and reach the surface of the planet, which is probably composed of a mixture of liquid hydrogen and helium. The descent and the long stay is protected by the Leidenfrost layer. The compounds resulting from the interactions rise to the surface within upward currents caused by the higher temperatures; they reach the upper atmosphere in the form of reddish-colored gases (probably of the same composition as the compounds in the belts, but more concentrated). So far, only two notable redspots have been observed on Jupiter, the small red spot that appeared in 1972 and disappeared about two years later (probably a relatively small comet) and the Great Red Spot with a life of at least three hundred years. The question that arises is:

can one attribute the cause of such a long life to a very large and highly dense *antimatter* mass? Naturally, this question cannot be answered at present, but the solution may come with the clarification of the natures of comets and cooled stars (Chapter VI).

According to these discussions, Jupiter may be regarded as a planet presently made up of a solid hydrogen core at approximately 0° K and at its almost maximum compressed state. Under these conditions, no heat is being generated within the center of the planet. Over the core is probably a thick liquid layer composed mostly of hydrogen and some helium and containing frozen compounds of heavier elements with as yet unknown compositions. On top of the liquid layer is the very deep atmosphere made up of hydrogen mixed with various quantities of helium, methane, ammonia and small quantities of water and inorganic and organic compounds of unknown nature (the by-products of +*matter-antimatter reactions*, some of which have reddish-brown color) which are being produced continually within the belts. Since our scant knowledge about the properties and behavior of solid, frozen hydrogen under very high pressure is mostly theoretical, we cannot at present determine with any degree of certainty either the diameter of the solid core, or the manner in which the rotation of the core produces the Jovian magnetic field.

The properties related to the evolution of the other three Jovian planets more or less follow those of Jupiter. The differences may be attributed to their smaller masses, their reception of small amounts of radiation from the Sun, their reduced chances of encounters with *antimatter* meteorites, and to the accumulation of lesser and lesser amounts of *antiH* atmospheres (in relation to their distances from the system's center).

We shall forego the discussion on the planet Pluto. The uncertainties about this planet make it difficult even to guess whether Pluto was originally a large satellite of the primordial Sun that had gone too far out and become surrounded by a +*H* atmosphere, or one of the large satellites of primordial Jupiter that accidentally encountered large amounts of *antiH* clouds. However, irrespective of its origin and whether it is made up of either +*matter* or *antimatter*, its probable high density indicates that it has been surrounded by a relatively dense antagonistic atmosphere which was instrumental in converting the satellite to a terrestrial-type planet.

Comets

Comets are spectacular objects with certain properties not shared by any of the planets or other satellites of the Sun. These may be enumerated briefly as:

1. When in the vicinity of the Sun, the comets are composed of round, compact and luminous nuclei within diffuse and nebulous comas, and followed by tails which can vary greatly in lengths.

2. On their way to the Sun, almost all comets begin to glow between the orbits of Saturn and Jupiter, with their luminosities increasing as they approach the Sun. The reverse takes place on the return leg of the journey; they become dimmer as they move away from the Sun.

3. The diffuse nebulosity that covers the nucleus is highly tenuous and grows in brightness as it approaches the Sun.

4. The size of the tail can be very small or extend as far as two hundred million miles; the tail is even more tenuous than the coma.

5. Spectroscopically, emission bands of unstable radicals and certain ionized molecules are emitted from the nucleus and the tail. However, if the comet comes very close to the Sun it emits lines of specific metals.

6. The orbits are seldom hyperbolic; they are usually either parabolic or elliptic; the two known instances of almost circular orbits are probably the result of perturbances by the planets.

We shall attempt to show these to be the properties of *antimatter* meteorites, visitors from outside the solar system, that are making random trajectories around the Sun in the course of their unpredictable movements within the Galaxy.

The prevalent theories ascribe the luminescence of the comets to fluorescence brought about by the recapture of electrons that have been knocked off by the ultraviolet radiation from the Sun.

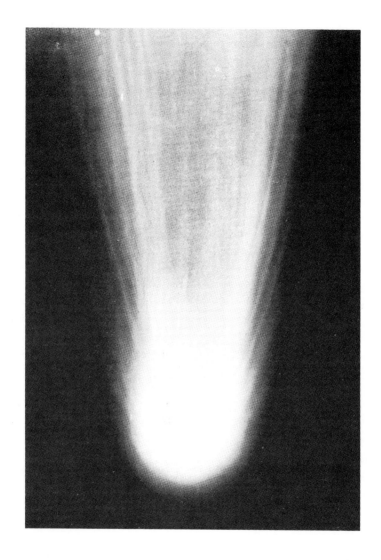

Figure 9. HEAD OF HALLEY'S COMET, MAY 8, 1910. *(Hale Observatories Photograph)*

There is nothing wrong with the theory, but the problem is that its provisions do not apply to the comets. The principal flaw may be found in the fact that ultraviolet rays do not have the power to penetrate deeply into matter and therefore cannot cause the incandescence of the fairly massive nucleus of the comet. The rays of the Sun did not make the "Mariner 10" satellite glow on its journey toward Mercury and around the Sun. Consequently, fluorescence cannot be regarded as the mechanism that makes the nucleus emit light. Some other explanation is needed for the cause of the luminescence of comets; the alternative seems to be the processes similar to the ones taking place on the surface of the Sun. Such processes can occur if we accept the nucleus as an *antimatter* mass which becomes luminescent when it passes through the *+matter* gases that surround the Sun.

At distances greater than a few astronomical units from the Sun, comets do not generate any light; whatever energy we may receive from them are the Sun's reflected light. But as they get closer and closer to the Sun they glow more and more. By the time they reach the orbit of the Earth they are completely luminescent and emit emission lines of C_2, OH, CN, NH and NH_2, all unstable radicals. We know the radicals decompose soon after their formation; therefore their presence implies their continuous synthesis by the reactions that are taking place around the nucleus. This means the source of energy is within the head of the comet. A more significant indication of the occurrence of *+matter-antimatter reactions* around the nucleus is the generation of emission lines of sodium, iron, magnesium, silicon and other metals by a comet if and when it passes very close to the Sun. This is an unexpected phenomenon and cannot be attributed entirely to the high temperatures resulting from the close proximity to the Sun. The reason for this statement is: sodium does not need very high temperatures to emit its spectrum, and if sodium is one of the common constituents of the nucleus we should receive its emission lines when the comet becomes fully aglow somewhere between the orbits of Earth and Venus and not only when it is very close to the Sun. The appearance of elemental sodium when the comet is in close proximity to the Sun implies that sodium and other metallic elements are created from *+H* when the nucleus comes in contact with relatively high concentrations of +*matter* atoms in the vicinity of the Sun. This type of formation of heavier elements is indicative of reactions similar to the ones that are taking place in the chromosphere.

The *antimatter* nature of the comets shows itself in another way, the manner of the increase in the intensity of luminescence as the comet races toward the Sun. The cause of the increase in luminosity may be ascribed to the continual increase in the rate of collisions made with the +*matter* atoms of the solar atmosphere. Two factors play parts in the increase in the rates of collisions: first, the nearer the comet is to the Sun the greater is its velocity and consequently the greater the number of encounters per unit of time; and second, the concentration of atoms increases as the distance from the Sun decreases, and hence the nearer the comet is to the Sun, the greater the number of contacts per unit of distance.

We have indicated the possibility of the solar atmosphere extending far beyond the orbit of Pluto—and the farther from the Sun the more rarefied the atmosphere. An approaching comet first passes through the extremely tenuous atmosphere; and though at this part of the journey it does not glow, it begins to heat up. After it passes the orbit of Saturn, the number of encounters increases, and this causes the temperature to become high enough for the nucleus to show the first signs of emission of light. Naturally, if the comet accidentally passes through a strong solar wind, the rate of heating is enhanced and it may be seen even before it reaches the orbit of Saturn.

The above statements are substantiated by the behavior of the Schwassmann-Wachmann comet, a comet which is traveling in an almost circular orbit between the planets Jupiter and Saturn. Because of its great distance from the Sun, this comet does not generate much light, but at times it brightens up to more than a hundred times its usual glow. The burst of luminescence has been traced to the periods when there are violent emissions of charged particles from the Sun. The probable explanation for the observed phenomenon is that the increased numbers of particles that collide with the comet cause a pronounced increase in the reactions on the surface of the *antimatter* nucleus, and this in turn results in the sudden intensification of light emission.

The particles that form the tails of the comets are most probably the by-products of reactions occurring within their comas, the diffuse nebulosities in front of the nuclei. The conditions in this region within the head of the comet approximately resemble those of the chromosphere and corona. But, whereas the Sun is extremely large and static relative to its atmosphere, the comets have very small masses and move at rapid rates through the solar atmosphere. These differences prevent the formation of confining radiation zones around the heads of the comets, but in return cause the formation of the tails. Because of the motion of the comet, the antagonistic atoms that collide with the front of the nucleus cause the flaking off of minute *antimatter* particles from the heated mass. The degree of the denseness of the head determines the rate of the flaking off and the sizes of the particles; these two factors in turn regulate the length of the tail. Once the particles become separated from the nucleus of a comet moving toward the Sun, they become vulnerable to the repulsive effects of the solar radiation and by slowing down fall behind the speeding head. But the tiny *antimatter* particles still have great velocities, and as they follow the nucleus they encounter $+matter$ atoms which keep them luminescent.

The radiation from the Sun is the principal factor impeding the motions of the particles; they gradually slow down, encounter a correspondingly smaller number of $+matter$ atoms and become proportionately dimmer in luminescence as their distances from the head increase. Finally, they lose the power of emitting light and become a part of the solar atmosphere. They float within the atmosphere until accidentally they encounter high concentrations of $+matter$ gases, such as the atmosphere of a planet. Then they become luminescent again, but this time they glow to total oblivion. We have records of such encounters. Earth has passed through the orbits of Biela (1872 and 1885) and Giacobini-Zinner (1946) comets, and each time it passed through the remnants of the tails spectacular showers of light were produced. These were not meteoric showers, but radiations produced by the annihilation of large numbers of microscopic *antimatter* particles that were trapped in the atmosphere of the Earth.

After the comets become observable, their orbits are found to be either parabolic or elliptic, but seldom hyperbolic. This has led to the belief that they are members of the solar system, as any visiting meteorite should have a hyperbolic orbit. This conclusion would be correct if the visiting meteorite did not encounter anything that would impede the rate of its motion. But the motions of comets are subject to continual interferences. The velocity of a comet at any point on its orbit is a function of two principal factors—the gravitational attraction from the Sun and the deceleration effect caused by the energy generated in front of its head. If the velocity were only a function of the gravitational attraction, an object entering the solar system from outside would follow the expected hyperbolic orbit. However, in the case of a comet, the radiation energy generated within its coma changes the trajectory. Thus, by the time a comet becomes observable, its course has already been changed to a parabola or an ellipse. Therefore, the calculated orbits of the comets are not indicatives of their real origins. They can originate from any point in the Galaxy and enter the solar system from any random direction.

One of the fundamental concepts in this book has been the principle of the existence of approximately equal amounts of +*matter* and *antimatter* and their more or less uniform distribution in the universe. Accordingly, we should expect equal numbers of +*matter* and *antimatter* meteorites to enter the solar system. This is probably what happens, but the difficulty of proof lies in the fact that the +*matter* meteorites cannot be easily observed. They do not glow and have very little possibility of colliding with Earth, where they can be identified from their compositions. In addition, the +*matter* meteorites have much better chances of leaving the solar system than the comets. They do not encounter much resistance from the solar atmospheric gases, they retain their hyperbolic trajectories, and they can escape from the solar system without encountering conditions that would betray their presence.

VI

THE EVOLUTION OF STARS

Most of the available information about stars comes from the Orion Arm of the Milky Way; the rest has been gathered from the observations of the globular clusters, the stars of the central region of the Galaxy, the Magellanic clouds, and our nearest neighbor, the Andromeda galaxy. The collected data point to one important fact: in certain regions of the Galaxy, stars share certain specific characteristics. For example, the stars of the globular clusters, numerous as they are, do not vary in size, color index and luminosity as do the stars of the arms. Furthermore, the stars of the nucleus have properties that are strictly their own. Yet, in spite of such diverse characteristics, the billions of stars that compose the Milky Way are still parts of one and the same evolutionary system, and the evolution of stars is intricately interwoven into the fabric of the evolution of the Galaxy. The two cannot be separated; consequently, the evolution of the stars should always be related to the evolution of the Galaxy.

The pattern of distribution of the evolutionary stages of stars within the Galaxy is very similar to that of the evolutionary stages of the living on Earth. In both cases the proper interpretation of the observed properties requires the knowledge of the position of the subject under investigation; with the living it is the geographical area and with stars it is their locations in the Galaxy. Accordingly, in the studies of stars we need the proper information about the general anatomy of spiral galaxies. The details of structures, classification and evolution of these galaxies may be found in Chapter IX. At this stage of the discussion we need only a simplified picture.

According to the accepted views, the Milky Way is in many ways similar to the Andromeda galaxy, which is a complete and mature spiral galaxy. If we take a close look at a fully evolved spiral galaxy of the Andromeda class, we find it to be composed of four major parts:

1. *The Nucleus.* An elliptically spheroidal structure in the center, it consists of densely packed old stars. The nucleus is indistinctly attached to the rest of the galaxy and in most photographs (e.g., Figure 10) it is not possible to distinguish the true boundary between the central structure and the arms to which it is attached. However, a general idea of the form of the nucleus may be obtained in some of the edge-on photographs (Figure 43) by extending the line of the curvature of the central bulge over the arms. The importance of this boundary lies in knowing the shape of the nucleus; this will be of help in determining the approximate shape of the embryonic nucleus at the time of the origin of the galaxy.

2. *The Primary Arms.* These are composed of fairly old stars which are compactly bound around the nucleus within the galactic plane. Because of their compactness they seem to be the continuation of the nucleus, but in the immature galaxies (Figure 40) they can be differentiated easily. The stars of these arms are younger than those of the nucleus but older than the stars of the secondary arms.

3. *The Secondary Arms.* These spectacular spiral arms can be seen without difficulty in most of the mature galaxies. They contain dust, interstellar gases and a variety of stars, the characteristics of which provide valuable information for the understanding of the pattern of stellar evolution.

4. *Groups of Independent Stars and Globular Clusters.* These do not follow the rotational motion of the galaxy, but move along independent orbits of their own around the nucleus. The stars in these groups have the characteristics of the older stars in the primary arms.

Ages of Stars

The estimation of the ages of stars is a necessary step in the determination of the status of their evolutionary stages. Among

Figure 10. RELATIVE POSITIONS OF THREE (OUT OF FOUR) PRINCIPAL
PARTS OF A MATURE SPIRAL GALAXY. NGC 3031 *(Hale Observatories Photograph)*

N—Nucleus   P—Primary Arms   S—Secondary Arms

all, the Sun is the only star whose age is known with a fair degree of accuracy. The methods used for the estimation of the ages of other stars of the Milky Way are subject to criticism. Most of the efforts in this line have been concentrated on determining the age from the rate of energy production, using the theory of thermonuclear fusion in the center of stars as the foundation for calculations. In particular, the rates of energy production of highly luminous stars have been used for the estimation of their ages. It has been argued that since the more massive stars generate energy at much greater rates, they are short-lived and hence are very young in age. This in turn has led to the concept of continuous birth of stars from interstellar gases and dusts.

Within the proposed theories of this book the lengths of the lives of stars are dependent on the masses of the *available atmospheric gases* which vary from star to star, even among the stars of the same mass. The salient idea in these new concepts is this: though the atmospheres of the stars were initially attained by the capture of antagonistic systems, in most cases the captured material has not been and will not be the only source of atmospheric substance; there is a continual replenishment (and even addition) of the antagonistic material as the stars tunnel their ways through the interstellar gas and dust clouds. *This implies that the stars with greater masses have better chances of living longer than the ones with smaller masses.* They can hold on to larger portions of the gases from their captured systems. Their intense radiations have converted all of the captured frozen antagonistic masses to a gaseous state (unlike the solar system where the Jovian planets are unusable); this provides them with proportionately greater fuel reservoirs. They can rob material from their smaller neighbors; and in competition with smaller stars, they can attract with greater forces the interstellar gases, dusts and nebular materials. Consequently, most of the larger and brighter stars, in spite of their prodigious rates of energy production, live longer than their smaller and thriftier brothers. *Hence, there is no reason to assume that the stars of the secondary arms of the Milky Way could not all be approximately of the same age.* This also means that we have to look to other methods for estimating the ages of stars.

In Chapter IV the general pattern for the formation of galaxies was discussed. In Chapter IX it will be shown that stars of the secondary arms of any spiral galaxy are approximately of the same age and the youngest in the whole system. These conclusions open the door for devising methods for the estimation of the ages of stars in different parts of the Milky Way. According to the theories of the evolution of galaxies in Chapter IX, the closer a region on the galactic plane of a spiral galaxy is to the nucleus, the older are the stars in that region. This empirical rule is only applicable to the stars of the galactic plane and does not apply to globular clusters and stars with independent orbits. But this rule does not give us the means for the accurate determination of ages of stars. We should also include the degree of compactness of stars in the region, for it indicates the rate of increase in their densities and hence their ages. When these and perhaps other factors are included in the estimations, we will be in a better position to determine the ages of stars with more precision. For the present we can only estimate relative ages according to their positions in the Galaxy.

Before proceeding with the discussion of stellar evolution, once again attention is drawn to the similarities between the evolutionary stages of stars in this chapter and those of galaxies and life described in later chapters. It will be seen that the phenomena of birth, growth, struggle for survival, dominance of the strong over the weak, old age and death are not restricted to the living. For those who want to attempt the exercise, they will not only find the attempts at correlating the evolutionary stages between the different systems extremely fascinating, but will also discover that by penetrating through the facades of nature they are able to see the true universe at its naked and not-too-mysterious form.

The First Evolutionary Stage — The Origin

All stars have begun their lives more or less in the same way as the Sun; the general pattern of a large, frozen, primordial system capturing a smaller frozen antagonistic system has been repeated billions and billions of time in galaxy after galaxy. Since there have been only two constituents to each star, the significant variation in the resulting characteristics have come about from the differences in the initial masses of the capturing and the captured systems. To put it more precisely, *the original mass of the capturing system determined the mass of the star and the ratio of the mass of the captured system to that of the capturing system was the prime factor in developing its properties.*

The Second Evolutionary Stage — The Adjustments

The stellar evolution cannot be dealt with in the same clear way as the evolution of the Sun. The solar system is an integrated system where each member has had a part in the formation and maintenance of its equilibrium. That is not the case with the diverse components of the stellar population. In an evolutionary system composed of large numbers of semi-independent members, we should expect deviations from the mainstream of the evolutionary trend by a small number of the constituency. But this should not distract us from following the true course of the evolution; at the same time attempts should be made to find the reasons for the out-of-the-ordinary patterns. In the evolution of stars, we consider the main sequence of the H-R diagram (Figure 11) as the principal evolutionary trend of the stars of the secondary arms. Accordingly, the giant-supergiant group, white dwarfs, Wolf-Rayet stars and other unusual stars are regarded as deviants that have strayed from the normal course by specific but explainable circumstances. The stars of the globular clusters, the primary arms and the nucleus are older stars that have already completed their evolutionary stages depicted by the classical H-R diagram. They will be discussed in their proper places later in this chapter.

The second evolutionary stages of individual stars have been very similar to those of the Sun. They consisted of adjustments between the antagonistic and the central systems. But, though in the stars of the main sequence the total masses of the captured antagonistic systems have been large enough to main-

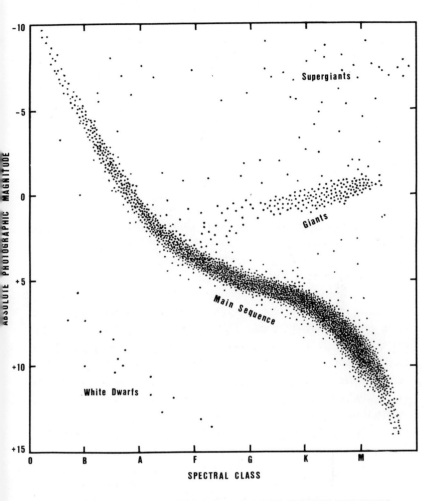

Figure 11. HERTZSPRUNG-RUSSELL DIAGRAM OF OBSERVABLE STARS

tain continual energy production at their maximum possible rates for long periods, this has not been the case with the giant-supergiant group. Many members of this latter group did not capture massive enough antagonistic systems to provide them with sufficient atmospheric material to produce energy at full capacity during the four to six billion years. Instead, the stars of the giant-supergiant group have gone through more evolutionary changes than the stars of the main sequence.

In contrast to the stars of the main sequence that reached their states of equilibrium a long time ago and have remained stable ever since, the members of the giant-supergiant group have had frequent changes of positions on the H-R diagram; these movements have been caused by variations in the concentrations of antagonistic atoms near their photospheres. In the beginning, most of them accumulated enough material to give them the capacity for a maximum rate of energy production, and as a result they joined the stars of the main sequence. But then the atmospheric gases began to deplete and the resulting reduction in their surface temperatures made them move to the right on the H-R diagram. The diminishing concentrations of the atmospheric atoms near the surfaces gradually weakened the radiation zones, which in turn brought about the lower surface temperatures and the expansions of their volumes. In some of them the radiation pressure dropped to such low levels that their internal materials could not be contained any longer; as a result the gases broke through the radiation barriers and expanded into space to produce some of the nebulae in the Galaxy.

Effects of Masses of Capturing and Captured Systems

The diagrams H and K of the Figure 12 are the diagrammatic reprsentations of the above statements. They are drawn to show the way the sizes of the masses of the capturing systems and the amounts of the captured antagonistic atmospheres have brought about the formation of most of the stars of the H-R diagram (white dwarfs and Wolf Rayet stars have not been included in the diagram—they are the products of later evolutionary stages and will be discussed in the next chapter). The five zones, A, B, C, D and E on Diagram K have been arbitrarily chosen and

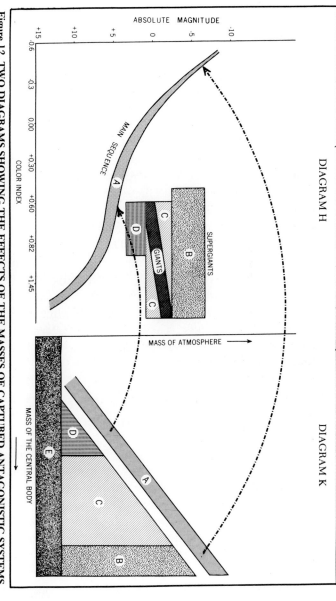

DIAGRAM H

DIAGRAM K

Figure 12. TWO DIAGRAMS SHOWING THE EFFECTS OF THE MASSES OF CAPTURED ANTAGONISTIC SYSTEMS IN RELATION TO THE MASSES OF CENTRAL BODIES DURING THE INITIAL STAGES OF EVOLUTION OF STARS. The coordinate "Mass of Atmosphere" represents the required minimum captured antagonistic mass for the frozen masses in zone A to become members of the main sequence; the greater the mass of the central body, the more massive should be the captured antagonistic system. The frozen masses in zones B, C, D, and E did not receive adequate antagonistic supplies and consequently became converted to supergiants, giants, subgiants and nebulae respectively. Diagram H: The H-R diagram of the stars of the Milky Way (principally members of the secondary arms). Diagram K: Diagram depicting the possible relationships between the masses of the central bodies and the masses of captured antagonistic systems in the newborn stars.

have no quantitative values. The zone A of the Diagram K represents proto-stars of different masses that received adequate supplies of atmospheric material and became converted to the stars of the main sequence. It is obvious that the greater the mass the more concentrated were the atmospheric gases around the central mass, hence the higher the resulting luminosity (larger volume and higher rate of energy production). *Thus during the transformation of the proto-stars of the zone A to the stars of the main sequence, the well-known relationship between mass and luminosity came into being.* The durability of this relationship confirms the conclusion that the states of equilibrium of the stars of the main sequence have been maintained at a steady rate for a long time. The zones B and C on Diagram K depict the proto-stars with large masses that captured antagonistic systems with smaller-than-adequate masses; they first joined the main sequence, but eventually ended up in the giant-supergiant zones B and C of the Diagram H. The masses of the proto-stars in the zone D, being smaller than the masses of the other two groups, were transformed to subgiants. The zone E portrays the proto-stars that captured very small antagonistic systems, enough to volatilize their contents, but not sufficient to convert them to stars. They became transformed to gaseous clouds, all of which most probably have been lost to the atmospheres of nearby stars.

The Third Evolutionary Stage — Progress Toward Stability

According to our understandings here, the state of stability of a system does not remain static, but is a gradual movement toward the theoretical limit of total stability. Accordingly, a galaxy is the state of equilibrium of multitudes of stars whose actions on each other have brought about a state of stability. The properties of stars within this unit are continually changing; within certain limits, these changes are improving the states of stability of the stars and hence of the Galaxy. To understand the manner of the evolution of stars, we have to learn about this movement, which is best demonstrated in the spiral galaxies where the stars of different ages are spread out inside the galactic plane for easy observation. Among these stars, those of the nucleus are the oldest and nearest to the optimum state of stability. Some have even passed that stage and have entered the dissolution phase of their evolutionary trend. The stars of the primary arms are not so old, but may be considered as being quite close to the maximum stable state. The members of the secondary arms, being relatively young, still have to go through a great deal of evolutionary adjustments. Therefore, our main goal should be to analyze and explain the properties of the stars of the secondary arms and then attempt to find connections between them and the older groups. In this way we can follow the life of stars from their infancy to their old age.

The H-R Diagram

The classical H-R diagram (Figure 11) is a representation of the contemporary state of evolution of stars of the secondary arms of the Milky Way. But since in the compilation of data for this diagram some of the members of other regions of the Galaxy have also been included, the diagram does not give a complete and true picture of the evolutionary positions of the stars of spiral arms. Still, this minor shortcoming is of little significance and the H-R diagram may be considered as the most reliable medium for the study of the stellar evolutionary trends. In the diagram, the stars of the main sequence belong to what we may call the "normal stars"—that is, those that have been generating energy at their full capacity throughout the lengths of their lives. However, before we embark on the comparison of the properties of these stars, we should point to a specific region on the main sequence which will be used in our studies later on. This area is located around the intersection between the +4.0 magnitude coordinate and the main sequence. Its significance lies in the fact that the H-R curve of the stars of M3 globular cluster makes a sharp turn in this location to join the main sequence (Figure 15-Diagram L). This is believed to be the place where a connection may be found between the younger and the older stars. Accordingly, in our discussions we shall divide the main sequence into two sections: the part with magnitudes higher than +4.0 will be referred to as the "high luminosity arm" and the portion containing the lower than +4.0 magnitude stars as the "low luminosity arm." It should be emphasized that the +4.0 magnitude is not a dividing line but an indication of an area of evolutionary significance.

Factors Influencing the Evolution of Stars

According to the hypotheses proposed in this book, five principal factors determine the course of the evolution of stars:

1. The concentration of the atmospheric atoms and dust particles in the vicinity of a star regulates the surface temperature; as a rule, the greater the mass of the star, the higher the concentration and the hotter the photosphere.

2. As long as a star has a sufficient reserve of atmospheric material, it produces energy at its maximum capacity and consequently its luminosity is a function of its mass. Most stars of the spiral arms are of this type and have had more or less fixed positions on the main sequence since they were formed.

3. The densities and the gravitational attractions of stars increase with age. As a result of this, their volumes shrink somewhat as they become older. But, in stars with adequate atmospheric reserves, the reduction in luminosity caused by the smaller volume is compensated by an increase in the higher concentration of atmospheric atoms and greater rate of energy production.

4. A star remains in its evolutionary position on the main sequence as long as it possesses sufficient amounts of atmospheric material; but when the depletion of the supply begins to show its effects on the rate of energy production, the surface temperature drops and the position of the star on the H-R diagram changes.

5. The smaller stars which belong to the "low luminosity arm," and are in competition with the more massive ones, are incapable of accumulating appreciable quantities of high-density antagonistic galactic dust particles for their atmospheres. Their reliance on the annihilation of lightweight gases (principally hydrogen and helium or their counter-equivalents if the atmosphere is composed of *antimatter*) for

their energy production is a factor in determining a different path of movement on the H-R diagram during their "evolutionary decline" than the stars of the "high luminosity arm." By "evolutionary decline" we mean the downward motions of the positions of the stars on the diagram after the depletion of the atmospheric material has begun to show its effects in lowering the surface temperatures and luminosities.

Evolution of Members of "Low-Luminosity Arm"

When the depletion of the atmospheric gases of a star belonging to the "low luminosity arm" begins to show its effects, both its surface temperature and luminosity drop; in other words, it moves downward along the main sequence. The important fact to remember is that this movement begins only after the star has used up most of its reserves and not before. Now, two stars of the same mass and on the same position on the main sequence do not necessarily possess the same amounts of atmospheric supplies; therefore, the one with the smaller amount is the first to move downward. This means that if we had accurate methods for the determination of the masses of a large number of stars of the "low luminosity arm," we would find that the mass-luminosity relationship does not hold for a number of stars in the lower section of the main sequence.

There is another aspect to the movement of the members of the "low luminosity arm" along the main sequence: when they reach far down and their surface temperatures drop to very low levels, they no longer generate light and cannot be observed and recognized as stars. Now, the smaller the mass of a star, the lower is its atmospheric reserve and it has correspondingly less chance of accumulating additional material from the Galaxy. Consequently, it has a shorter life than the more massive ones. Since all the stars of the secondary arms came into existence approximately at the same time, by now the stars with smaller masses and shorter lives have already cooled down and no longer generate light. *Therefore, in the present state of stellar evolution, there is a minimum limit for the mass of stars emitting light.* The stars with masses below that limit are not shining any more, but similar to all the members of the secondary arms, they are still orbiting around the nucleus on the galactic plane. We have evidence of the existence of these cooled and dense stars from their silhouettes against bright nebular backgrounds (Figure 13). In the literature they are referred to as "globules."

If with the process of aging, the stars of the "low luminosity arm" eventually move downward on the main sequence and lose their luminescence, we should expect the future masses of the smallest observable stars to be larger than the smallest contemporary stars. This is probably the reason for globular clusters and the primary arms not having stars with masses comparable to the smaller stars of the spiral arms. We should also expect a change in the future shape of the "low luminosity arm"; by the time the stars of the present secondary arms reach the age of the contemporary globular clusters, the "low luminosity arm" should look somewhat similar to the lower section of the H-R diagram of M3 stars (Figure 15)—that is, shorter than the present length and wider.

Evolution of Members of "High-Luminosity Arm"

Indications for the direction of the evolutionary movements of the luminous stars belonging to this arm on the H-R diagram come from two sources, the H-R diagram of the stars of galactic clusters (Figure 14) and the superimposed H-R diagram of the M3 stars on the main sequence (Figure 15-Diagram L).

The Figure 14 is comprised of the H-R diagrams of a number of galactic clusters and one globular cluster. A galactic cluster, unlike the globular one, consists of a group of relatively young stars in the transient state of stellar association inside the secondary arms of the Milky Way. This diagram predicts with a fair degree of accuracy the effects of aging on the direction of evolutionary movements of high luminosity stars of the main sequence. Most of the stars of the galactic clusters in Figure 14 belong to the "high luminosity arm." The diagram consists of a base curve, an amalgamation of the lower luminosity stars of the clusters, from where a number of offshoots branch out toward the low temperature zone. Each branch represents a

Figure 13. THE LAGOON NEBULA IN SAGITARRIUS *(Official U.S. Naval Observatory Photograph).* The black globules are most probably cooled red dwarfs silhouetted against the bright nebula.

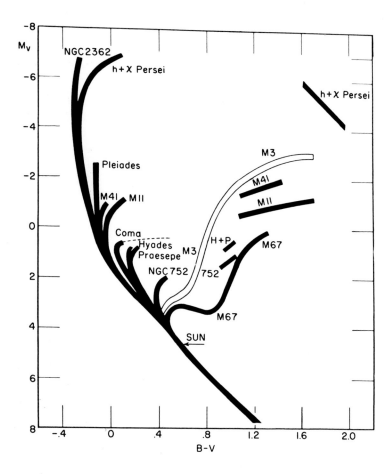

Figure 14. THE H-R DIAGRAM OF THE STARS OF SEVERAL GALACTIC CLUSTERS AND ONE GLOBULAR CLUSTER (M3). By A. Sandage: *Astrophysical Journal*, 125:436 (1957).

cluster and contains a number of the brightest stars in that group. With the exception of the Pleiades cluster, all the off-shoots curve toward the right side, indicating that the greater the luminosity of a star in that branch the lower its surface temperature. The nonconformance of the Pleiades cluster is most probably caused by the shifting of wavelengths toward blue by the bright nebulae covering the stars and consequently may be kept out of the discussion.

The two interesting parts of the diagram are: first, it is only the brightest stars in each cluster that have lower surface temperatures, and second, this relationship is specific to each cluster without regard to the luminosities of the stars of other clusters. In other words, although the lower-luminosity stars of one cluster on the base curve may have magnitudes higher than the brightest stars of a second cluster, the brightest stars of the second one, irrespective of their luminosities, are on branches that are curving toward the lower temperature zone. For example, in the h and x Persei cluster the stars with magnitudes higher than -4 are the only ones on the curved branch, and the less luminous ones are on the base curve. Meanwhile, the brightest stars of the M11 cluster with magnitudes between -1 and +1 (values that are lower than the magnitudes of stars of the h and x Persei on the base curve) form a curving branch of their own. This is not an isolated case and occurs repeatedly on the diagram. The inference is clear: *the lower surface tempera-tures in relation to the masses of the stars within any specific cluster is a regional phenomenon related to the atmospheres of the stars within that cluster.*

If the rate of annihilation of matter relative to surface area is the prime factor in determining the surface temperature of a star, then the more massive stars in each cluster do not possess sufficient atmospheric reserves to produce energies at their full capacities. In the previous paragraph it was concluded that the lowering of the temperatures of the largest stars in each cluster was a regional phenomenon. A galactic cluster is an association of stars within a region inside the Galaxy; this limits their sup-plies of antagonistic gases. During their formation, most prob-ably each star received its share of the atmospheric material on the basis of its mass. Also, whenever a cluster passed through a dust cloud each star received its share. This type of division of acquired material may be seen in its exaggerated form in the distribution of nebulosities between the stars of the Pleiades cluster (Figure 28). Accordingly, the cause of the reduction of the surface temperatures of the more massive stars within galactic clusters may be ascribed to the depletion of atmospheric gases brought about by their relatively rapid rates of consumption of the limited supplies within the confines of clusters.

If the above interpretations are correct, then the features of the Figure 14 diagram are pointing to the direction of the evolutionary motions of the bright stars of the main sequence on the H-R diagram. The direction is really the same as that of the stars of the "low luminosity arm"—that is, downward and toward lower temperatures. But since the rates of declines are different, the direction of the movements are more exaggerated. However, as we shall see later, the survival of the high-luminosity stars is conditional to their being able to accumulate dust particles from space. Being massive, most of them will not have difficulties in adding these heavy and dense atoms to their atmospheres. But there will be some that would not encounter dust clouds and cannot accumulate sufficient quantities of heavy particles; these would most probably slide down along the main sequence as their atmospheres become depleted.

In addition to the curves of the galactic clusters, the Figure 14 diagram also furnishes a curve for the much older stars of the globular cluster M3. On examining the entire diagram, one notices that all the branches containing the brightest stars of the galactic clusters (with the exception of Pleiades) are pointing toward the curve of the M3 cluster. These curved offshoots are the product of four to six billion years of aging. During this period a trend has been established, and consequently, as the stars grow older, we should expect the extension of the branches to be in the direction of the M3 curve. Now, if we assume that the majority of bright stars of the main sequence will also follow the pattern of movements of the stars of the galactic clusters, then the M3 curve portrays the approximate positions of the stars of the "high luminosity arm" when they reach the present age of the M3 cluster.

The lowest galactic cluster branch on Figure 14 diagram belongs to the M67 cluster, a rather crowded cluster. It has produced a curve similar to that of the much older stars of M3, thus substantiating the supposition that the relationship between the mass of the star and the available atmospheric material is the prime factor in determining the position of a star on the H-R diagram. In the M67 cluster, each star draws its atmospheric material from a comparatively small space. This is a typical picture of the effects of overcrowding among stars. Though the compositions of their atmospheres are different, the relatively young stars of the M67 cluster are deprived of the atmospheric material in the same way as the older stars of the M3 cluster; hence the similarity of the two curves.

Evolutionary Pattern of Stars of Secondary Arms

If we accept the H-R diagram of the M3 cluster as that of aged stars, then by superimposing it on the classic H-R diagram we can learn about the future course of the evolution of the stars of the secondary arms. The diagram L of the Figure 15 depicts this superimposition; the combination of the two curves indicates the expected direction of the movements of the brighter stars of the main sequence as they grow older (the smaller red dwarfs would have cooled down by then). The shaded Y-shaped curve of the M3 cluster consists of two principal parts: first, the major branch which starts at the extreme right, curves downward to the left and on crossing the main sequence bends sharply to the right; and second, the minor branch which is attached to the main curve at the Y junction and extends horizontally in the proximity of the magnitude 0 coordinate. The major branch is believed to represent the brighter stars of the main sequence when they reach the age of the present globular clusters. The minor branch probably consists of the aged supergiants of the globular cluster and hence is a guide for determining the possible future positions of the supergiants of the secondary arms.

On the basis of the indications given by the stars of the galactic clusters, arrows are drawn on Diagram L to show the expected direction of the movements that the high-luminosity stars of the main sequence curve will take as they age. The arrows in-

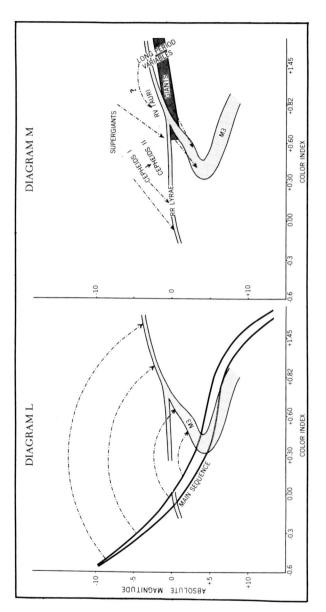

Figure 15. EXPECTED PATTERN OF MOVEMENTS OF THE STARS OF THE H-R DIAGRAM AS THEY GROW OLD.
Diagram L: Superimposition of the H-R diagram of the M3 stars over the main sequence. The arrows show the expected directions of movements of the stars of "high luminosity arm" as they age. Diagram M: The expected directions of movements of giants and supergiants of the spiral arms on the H-R Diagram as they approach the age of M3 stars.

dicate only directions and not necessarily the paths. Also, most probably the ultimate destination of the stars of the "high luminosity arm" is not the upper portion of the major branch of the M3 curve but beyond it. In fact, there is a great possibility that as the stars age and the main sequence curve bends, the movement will not stop at the present position of the M3 curve but pass it, probably as far as the M67 curve in Figure 14. However, it is expected that the downward movement will eventually come to a halt and its direction be reversed. This change in the direction, which most likely has already taken place with the curve of the M3 stars, may be attributed to the gradual and fresh accumulation of high density atmospheric atoms from the bodies of the cooled stars in the cluster. We shall discuss this aspect of evolution later on.

The bent arm of the major branch of the M3 curve almost overlays the "low luminosity arm" of the main sequence curve, implying that the low-luminosity stars of the globular clusters possess approximately the same luminosity-to-surface-temperature relationship as the younger stars of the spiral arms. But there is a big difference between the two groups: the stars of the M3 cluster are older and denser than the population I stars, and hence, for the same luminosity values, possess correspondingly lower atmospheric reserves. When we include the densities of the atmospheric atoms of the M3 stars (as we shall see presently, they are denser and heavier than the normal atmospheric atoms), we come to the conclusion that these older stars are starved for atmospheric material and are not producing energies at their full capacities. This means that the low-luminosity stars of M3 are more massive than their luminosity-surface-temperature equivalents on the main sequence.

There are two principal reasons for the belief that the stars of the minor branch of the Y-shaped M3 curve are the old supergiants of the globular cluster. The more important one is the tendency of the supergiants of the spiral arms to move downward and to the left as they age. This direction of movement may be ascribed to the gradual increases in the gravitational attractions between the interior atoms which cause reductions in the volumes of the huge stars and consequently lower their luminosities while increasing their surface temperatures. The latter change is brought about by the more powerful gravitational forces attracting greater numbers of atoms, especially the dust particles, and concentrating them near their photospheres. The second reason, which is merely ancillary to the first one, is the presence of the RR Lyrae stars in a specific section of the curve. It will be shown presently that these pulsating stars are probably the old Cepheid variables of the globular cluster. The RR Lyrae group have much shorter pulsating periods than the Cepheids. But, as will be seen in the next section, an increase in the gravitational attraction of the atoms inside the stars proportionately reduces their rates of pulsation.

Evolutionary Pattern of the Old Stars

The next question that arises is: what is the expected pattern of the H-R diagram when the stars are older than the present age of the M3 cluster? The obvious place to look for the answer is the H-R diagram of the stars in the nucleus. The available information is unfortunately sketchy but can be used. By the joint application of the principles in the proposed hypotheses and the known properties of the stars of the nucleus, we can cautiously predict the shape of the H-R diagram for the surviving stars in the last stages of their evolution. But before doing that we have to digress and discuss the changes that take place in the nature of the stellar atmospheres as they grow old.

One of the conclusions drawn from the continual increase in the strengths of gravitational forces of stars as they grow old is that they bring about the folding and the contraction of the secondary arms of the Galaxy (for details see Chapter IX). This means that as the distances between the stars become reduced, more and more stars enter the nebular and dust clouds and get the chance to incorporate some of the heavier dust particles into their atmospheres. The incorporation occurs if the dust clouds and the atmospheres are composed of the same type of matter; if not, the dust clouds do not mix with the atmospheric gases and remain separated until they encounter the right type of atmospheres of other stars. Therefore, one of the changes associated with aging is the gradual disappearance of dust clouds. This is evidenced by the fact that the older a region within a galaxy is, the fewer dust clouds it contains. It also provides the reason for the presence of a plentiful supply of dust clouds in the spiral arms of galaxies, their scarcity within primary arms, and their absence inside nuclei.

The secondary arms of our Galaxy contain a large number of small stars, most of which are members of the "low luminosity arm" of the main sequence and go under the general term of red dwarfs. As already mentioned, eventually the atmospheres of these stars will become exhausted and they will turn into cold and dense "globules." The third stage of the evolutionary life of these stars has been completed and they are ready for the fourth stage, their dissolution. Although after the loss of their atmospheres, the densities of these cold masses remain the same, they will be attracted to and captured by the more massive stars whose gravitational attractions are on the increase. They will become satellites of the larger stars and their dissolution will consist of their step-by-step conversion to the gaseous state which is then used for the replenishment of the diminishing reserves of their captors.

Depending on the composition of a cold star relative to the nature of the atmospheric material of its captor, the dissolution of the captured star follows one of these two courses:

1. If the newly acquired satellite is composed of the same type of matter as the star's atmosphere, it becomes subject to the effects of the increase in the gravitational attraction. As it orbits around its captor, it is pulled in very slowly, and when close enough it begins to vaporize. It becomes converted to high density gases which are then incorporated into the atmosphere and used for energy production.

2. If the satellite is made up of the same type of material as the central mass and is antagonistic to its atmosphere, it heats up again and becomes transformed to a star, forming a binary with its captor. Here, two stars will be using the same atmospheric material and consequently will have relatively short lives. After they cool down, they can no longer protect themselves by keeping their distances from the larger stars through their radiation energies and are eventually captured. The cycle then repeats itself and both become incorporated into the atmospheres of other stars.

We can now draw certain conclusions regarding the last stages of the evolution of stars as represented by the H-R diagram. The Figure 15 diagrams indicate that by the time the stars of the contemporary secondary arms reach the present age of globular clusters, the H-R diagram would be somewhat similar to that of the M3 curve. As the stars grow older the curve will break into two parts at the pivot point where it bends sharply. The stars in the lower section with magnitudes lower than +3 to +4 will move downward along the main sequence channel and turn into cold stars. On the other hand, the high-luminosity stars will move upward haphazardly into a shapeless area above the horizontal arm of the M3 curve; the rise in luminosity will be due to the continual addition of gases secured from the bodies of the cooled stars. This stage will take a long time without any definite or abrupt end. During this period, as stars die and become assimilated by others, the formless area on the H-R diagram will slowly fade away. Most probably these changes will be coinciding with the dissolution of the Galaxy itself.

Pulsating Stars

This presentation of evolution of stars would be incomplete without including the variables. However, since this group of stars compose a very small percentage of stellar population, they cannot be considered as important evolutionary links. Their significance—besides their usefulness as astronomical tools for measurements of distances and luminosities—lies in their unusual rhythmic behaviors. Now, whether a star has a steady existence or has its peculiarities, it is still a star that generates energy; consequently, the theories of energy production for stars should also be applicable to them. Accordingly, explanations should be provided for the formation of pulsating stars.

Almost all the pulsating stars belong to the giant-supergiant class. This necessitates a review of the properties of this group of stars and the mechanics of their energy production with the aim of providing the reasons for the causes and mechanisms of pulsation among some of its members.

The giant-supergiant group are luminous, massive, tenuous, voluminous stars with surface temperatures lower than their mass equivalents on the main sequence. The high luminosities are due to their large volumes which provide them with extensive radiating surfaces. All of their described properties may be attributed to insufficient amounts of atmospheric material which produce relatively low radiation pressures. They hold their forms through the joint effects of the gravitational attraction of relatively dense atoms (the spiral arms members have had 4-6 billion years to build up their densities at rather slow rates) and the confining effects of the weak radiation pressures surrounding them. These properties do not provide protection from outside intruders; such objects as dense cold stars, meteorites and meteors can gain entry into their interiors and interfere with their steady rates of energy production.

In contrast to the bright stars of the main sequence, which are protected from intruders by the effects of their powerful radiation zones and high surface temperatures, the members of giant-supergiant group are vulnerable to these intruders' entry. They are massive and can attract cooled stars and other objects; they are voluminous and provide large areas to collide with; their atmospheres are not sufficiently dense to prevent some of the objects that collide with them from passing through to the interior; and their surface temperatures are not high enough to vaporize the visitors before they can penetrate deep into the body of the star. Consequently, the probability of large and dense objects reaching and entering the star is far greater for the members of this group than with their mass equivalents on the main sequence. Here we are interested only in intruding objects composed of material that is antagonistic to the atoms in the interior of these stars.

Cause of Pulsation

Usually, the captured objects do not collide with the stars after they come under their gravitational effects, but on occasion their paths cross and a collision occurs, in which case the intruder plunges into the massive star. If the object is small, it does not have a lasting effect; but if it is relatively massive, dense, and antagonistic to the internal gases it easily passes through the atmosphere (which is of the same type as the intruder) and the radiation zone, and penetrates into the interior. The intruder does not explode, as the tenuous gases inside the star produce a radiation zone around the solid mass and protect it from very rapid evaporation. In the first part of its inward journey the intruder heats up at a relatively slow rate; this provides time for the object to penetrate deep into the interior. However, this movement, together with the rather quick rate of expansion of the object as it becomes hot and gaseous, set up a series of turbulences which cause the eventual mixing up of the antagonistic gases and the production of tremendous amounts of energy. Most probably the reactions do not cause an explosion, but they do bring about the generation of great quantities of energy at rapid but orderly rates. As a result the internal gases rush outward.

The onrushing atoms penetrate into the rarefied atmosphere, bringing about increases in + *matter-antimatter reactions* on the surface and steadfastly strengthening the radiation pressures which when powerful enough will put a stop to further expansion. The end of the expansion comes after the volume has become much greater than the normal size of the star. Similar to all phenomena of this type, overexpansion takes place and is then corrected by contraction. The momentum carries it beyond the point of equilibrium (the median point) and creates a state of overcontraction. The cycle then repeats itself again and again, turning the voluminous star into a pulsating one.

Pulsation Mechanism

The four curves on the diagram of Figure 16 are drawn to relate the brightness, surface temperature, radial velocity and volume changes in the course of pulsation of a typical cepheid variable, Delta Cephei. The following relationships may be noted:

1. Maximum light production corresponds with highest surface temperatures; minimum brightness is associated with lowest surface temperatures.

2. The star is hottest and brightest at about its maximum rate

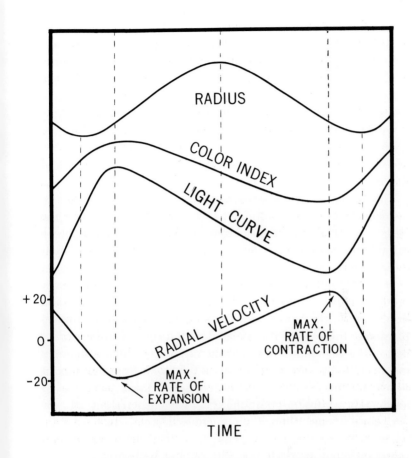

Figure 16. **RELATIONSHIPS BETWEEN THE PATTERNS OF BRIGHTNESS, SURFACE TEMPERATURES, RADIAL VELOCITIES AND RADIUS OF DELTA CEPHEI.** The maximum rate of expansion occurs shortly after the completion of contraction. The maximum rate of contraction takes place before the volume reaches its minimum size.

of expansion—that is, when the radiating surface gases are rushing outward at their maximum speeds. This occurs soon after the star has passed through its minimum volume phase.
3. The lowest brightness and the coolest stage occurs approximately at the time of maximum rate of contraction. These two cycles then change directions together before the volume reaches its minimum size. This is also the beginning of acceleration in the rate of energy production.

We shall now attempt to explain these relationships in terms of the proposed hypotheses.

The maximum rate of expansion occurs when, in response to overcompression, the volume has begun to expand and the highest possible number of interior atoms surge outward, break through the radiation barrier and become mixed up with the atmospheric gases. This condition results in the maximum amount of *+matter-antimatter reactions* and the highest rate of energy production (indicated by maximum surface temperatures and brightness on the curve). But since the continuation of the expansion of the volume is associated with a corresponding reduction in the velocities of the peripheral atoms, with the enlargement of volume, fewer and fewer of the atoms from the interior penetrate through the radiation zone and the rate of energy production drops proportionately. The penetration of the peripheral atoms, though on a reduced scale, does not stop when the volume reaches its peak; the rate of energy production continues along its declining pattern, but without any abrupt drop.

The beginning of volume contraction is not associated with any significant interruption in the *+matter-antimatter reactions*: at those high temperatures the velocities of atoms are high, and though on the average their motions are toward contraction, many of them do pass through the much weakened radiation zone. As the contraction progresses, the atmospheric gases that are following are kept back by radiation, and since there is no pressure from the outside to force the gases inward, the two antagonistic gases keep separating more and more and the energy production continues to drop. Finally, the rate of contraction reaches its maximum; this is when the central core has become quite congested and its gravitational forces are pulling in the peripheral atoms (including the atmospheric ones) at the fastest possible rate. Consequently, the atmospheric atoms penetrate through the radiation barrier and into the crowded region to become annihilated at an increasing rate. That is when the rate of energy production begins to accelerate and the light and temperature curves change directions and turn upward. As the volume continues to decrease toward its minimum size, the energy production continues to increase in intensity, with the maximum rate coinciding with the maximum rate of expansion, which occurs after the volume has begun to increase in size.

Density Pulsation-Rate Relationship

If the pulsation of a star is the product of intermittent penetration of atoms into the antagonistic regions and the amount of produced energy is a function of the number of atoms involved in these incursions, then the greater the densities of atoms in a pulsating star, the greater the rate of energy production; hence the period of pulsation should be shorter. This relationship has been put in mathematical terms where it has been shown that in pulsating stars with regular cycles the rate of pulsation follows the equation:

$$P^2 d = k$$

where P is the pulsation period, d the mean density and k a constant. This means that the more tenuous a star is the longer is its period of pulsation, with the rate decreasing in proportion to the square root of the mean density of that star. We shall make use of this relationship in the following discussions, but first let us review briefly the general classification of the pulsating stars.

Classification of Pulsating Stars

Just over 8200 pulsating stars have been observed and recorded. Table I gives their classification, distribution in the

Galaxy and the approximate percentage of each group; the last may be used as a rough measure of the relative abundance of the different classes. According to the theories in this book, the positions of the variables in the Galaxy should provide us with their approximate relative ages. This information gives us the means to estimate the direction of their movements on the H-R diagram as they grow older. However, this is not sufficient grounds on which to establish an evolutionary trend. The conclusion should also be supported by such information as the effects of aging on the spectrum characteristics and the influences of densities on the rates of pulsation.

In discussing the properties of the various classes, we shall begin with the semi-regular and irregular groups of variables. The principal criteria used for the grouping of these variables in one class has been their irregular plusation characteristics, and no consideration has been given to their distribution within the Galaxy. The class consists of variables from the central region, the halo and the spiral arms: in other words, a conglomerate of stars of various ages. Consequently, the group is of little value in the evolutionary studies. Nevertheless, their irregular behaviors may be used to advantage in substantiating the proposed theories regarding the causes of pulsations.

The irregularities of pulsation in these variables may be ascribed either to multiple collisions with solid, antagonistic masses, or if only one large mass has entered the star, to the location in the interior where the maximum energy was generated. In the first case, a number of collisions at different times might have caused interference patterns between the expansion and contraction movements originating from a number of sources within the sphere. In the second case, where the position of maximum energy production was at some distance from the center, the uneven spread of the expansion and contraction movements could have brought about the irregular patterns; here, the crests of the movements would not reach the various locations on the surface at approximately the same time, and thus the back-and-forth expansion and contraction movements could interfere with each other to produce at times constructive and at other times destructive effects. It is interesting to note that all the semi-regular variables are M giants and supergiants

TABLE I. THE GENERAL PROPERTIES OF IMPORTANT GROUPS OF PULSATING VARIABLES

Type	Characteristics Spectrum	Period Days	Description	Distribution	Approximate percentage of all known variables
Relatively young variables					
Cepheids—type I	F to G supergiants	3 to 50	Regular pulsation	Spiral arms	6.1
Long period (Mira-type)	M red giants	80 to 600	Almost regular pulsation	Spiral arms	36.9
Beta Canis Majoris	B giants	0.1 to 0.3	Maximum light at the time of highest compression	Spiral arms	0.1
Older variables					
Cepheids—type II	F to G supergiants	5 to 30	Regular pulsation	Halo and near nucleus	0.5
RV Tauri	G to K giants	30 to 150	Alternate large and small maxima	Halo and near nucleus	0.9
RR Lyrae	A to F giants	less than 1	Regular pulsation	Halo and near nucleus	24.5
Mixture of variables of all ages					
Semi-regular Irregular	All types giants and supergiants	30 to 2000	Periodicity not dependable	Spiral arms, halo and nucleus	30.7

that possess tenuous interiors and are surrounded with abnormally rarefied atmospheres. This type of structure has a much greater chance of developing irregularities in its pulsations; it is large in size, has little internal cohesion and is likely to develop interference patterns. However, one fact emerges from these discussions: the mere nature of irregularity is an indication of transitory nature of this type of pulsation; the interferences will eventually bring about the cancellation of expansions and contraction. Unless new collisions occur, the stars will probably gain back their more stable states and turn gradually into non-pulsating forms.

The next group to be considered is the relatively young variables of the spiral arms. Three classes are in this group: the long-period group of Mira-type, the type I Cepheids and the Beta Canis Majoris stars. Since the members of these three groups are approximately the same age, their differences are most probably caused by their sizes, their densities and the intensities of energy production on their surfaces. The following provides a resume of the pertinent data about these stars:

Type	Spectrum Characteristics	Period
Long-period of Mira-type	M red giants	80-600 days
Cepheids type I	F to G supergiants	3-50 days
Beta Canis Majoris	B giants	0.1 to 0.3 days

In these types a certain relationship may be noticed between their surface temperatures and their periodicities: on the average, the higher the surface temperature the shorter the period. We may arrive at this empirical relationship in another way. It has been stipulated that the increase in the density of a star is a function of its age and its rate of energy production. Since the stars of the spiral arms are approximately the same age, their densities can be considered as functions of their surface temperatures. By combining this temperature-density relationship with the density-periodicity formula, we arrive at the above-mentioned relationship—that is, the temperature of a relatively young pulsating star is a rough measure of its periodicity and vice versa.

This interpretation of the above three related properties in terms of the proposed theories leads to the conclusion that the concentration and the nature of the atmospheres of the variables are the fundamental factors in bringing about their observed properties. We may then go a step farther and suggest the following:

1. The long-period giants have had relatively meager amounts of atmospheric material throughout their lives.

2. The type I Cepheids started either with low concentrations of atmospheric atoms and increased their supplies by acquisition of dust particles, or they had acquired medium quantities of material right from the beginning and have more or less kept up with the level by collecting new material.

3. The Beta Canis Majoris group began their lives with a low or medium concentration of atmospheric material (at that time they were M to G giants); at this early stage they collided with cold, antagonistic masses which turned them into pulsating stars; later on they accumulated high concentrations of atmospheric material from dust clouds with rich supplies of antagonistic dust particles and turned into B pulsating stars.

It should be remembered that all the above three types are very massive stars which, in order to generate energy at their full capacities, require large amounts of atmospheric material. There are other points about the evolutionary states of these stars which will be discussed shortly.

Significance of the Variables' Characteristics

At this juncture we shall digress to review briefly some of the pertinent points regarding the characteristics of the pulsating stars. The following is the summary:

1. The expected evolutionary motions of the majority of the members of giant-supergiant group (including the variables) on the H-R diagram is from right to left and downward. This conclusion is based on the expectations that most of these stars

will continue to accumulate sufficient amounts of atmospheric material while their volumes are contracting; but there is no certainty that all of them will get the chance to replenish their consumed atmospheric atoms as they grow older. In case of such deprivation, until the increased densities and the accompanying reductions in volumes show their effects, the movements on the H-R diagram will be in the same direction as those of the stars of the galactic clusters, that is, from left to right and downward.

2. Among the contemporary pulsating stars of the primary arms, with the directions of movements from left to right and downward (e.g., the type II Cepheids), the surface temperature-periodicity relationship does not hold as closely as for the variables in the spiral arms. If the rates of energy production of these stars have been greater in the past than at present, then their densities are higher and their periods correspondingly shorter than what their contemporary temperatures would indicate. The temperature-periodicity relationship has been arrived at by inter-relating the density-periodicity and temperature-density relationships. Consequently, if the present temperatures do not correspond with the true values of the densities under consideration, then the reliability of the temperature-density relationship becomes questionable for this group of variables.

3. It has been mentioned that unless the intruders penetrate deep into the huge stars, regular pulsation will not occur. If this be the case, then the higher the surface temperature of the star, the greater the sizes of the intruding objects should be. It also implies that no matter how large the size of the object, it cannot pass through the barriers of high temperatures, intense radiation pressures, and the relatively dense and deep atmospheres of the stars that have surface temperatures above certain values. All the relatively young pulsating stars of the spiral arms (with the exception of Beta Canis Majoris group) are cooler than the F classification. This seems to be the highest temperature limit; at higher temperatures the intruders most probably become vaporized before they can reach the interiors.

4. The Beta Canis Majoris Class B group, composing about 0.1% of the known variables and 0.25% of the regular pulsating stars of the spiral arms, seems to be an exception to the above explanations. But, as it has already been indicated, at the time of their collisions the members of this small group were most probably much cooler than at present. It was after their conversion to pulsating stars that they came across relatively rich dust clouds and became hotter and moved closer to the main sequence. The support for this belief comes from the rates of rotation of these stars; they rotate at much slower rates than their non-pulsating equivalents on the main sequence. This difference indicates that the two groups have arrived at approximately the same destination via different evolutionary routes.

5. The presently observed rate of rotation of any star is an indicator of the evolutionary course it has followed. For example, in the solar system the entanglements of the satellites of the two primordial systems caused the loss of the Sun's original angular momentum; this in turn produced a type of stellar rotation which is probably typical of a number of planet-bearing stars of the low-luminosity arm of the main sequence. In contrast, the original bright stars of the H-R diagram have retained a fair portion of their angular momentums. As mentioned in Chapter V, these stars began their lives with very massive frozen central bodies surrounded by large satellites orbiting very closely to their central masses. When a system of that formation captured an antagonistic system, these satellites became a part of the central body at a relatively fast rate and carried with them their angular momentums—and there were no planets to interfere. As a result, most of the angular momentum was preserved. On the other hand, if the captured antagonistic system was not large enough for a very massive frozen body to provide it with the needed atmospheric gases, as was the case with the members of the giant-supergiant group, then the resulting low surface temperature and tenuous interior prevented the star from retaining most of its angular momentum. The consequence was a slow rotating star.

6. The rates of rotation of the stars of the Beta Canis Majoris group, relative to their equivalents on the main sequence, point to the possibility of differentiating the bright stars that directly joined the main sequence from those that reached it via the giant-supergiant route. The high-luminosity stars with rates of rotation faster than those of the Beta Canis Majoris stars are most probably the ones that became members of the main sequence right from the beginning. Those with rotation velocities comparable to the rotations of these variables are the giant-supergiants that acquired rich supplies of atmospheric material later on.

The Evolutionary Trends of the Spiral Arms' Variables

Returning to the discussions of the evolutionary trends of pulsating stars, the diagram M of Figure 15 gives the expected courses for most members of the two prominent groups of variables in the spiral arms. This suggested pattern of movements on the H-R diagram is based on the already outlined principles for the evolution of the giant-supergiant group—that is, they will continue to increase their densities, and this in turn will bring about a gradual reduction in luminosities and higher values for surface temperatures.

Though empirical in nature, the relative abundance of the different groups of pulsating stars can become of help in showing certain evolutionary possibilities. The limitations of this approach come from the fact that the population densities of stars in the neighborhood of each pulsating star contribute to its evolution, and these are the states about which we have very little knowledge. For example, the stars of globular clusters are under different sets of conditions than the ones in the primary arms. Both groups are approximately the same age, but the stars of globular clusters are subject to greater space restrictions than the ones in the primary arms. In spite of these uncertainties, the relative abundance of certain groups of variables can be informative in special instances.

One such case involves the Mira-type variables. In the diagram, their evolution has been questionably directed toward RV Tauri-type stars. Now, whether or not some of the long-period variables become transformed to RV Tauri stars as they grow old, the fact remains that the relative abundance of these two groups does not substantiate this evolutionary path for the majority of them. The Mira-type variables compose about 36.9% of all known pulsating stars as compared to 0.9% of the older RV Tauri group. On the H-R diagram there are no other regularly pulsating old stars to the left of these M pulsating giants. The only alternative then is to accept that aging causes reconversion of most of these variables to their original non-pulsating forms, which then follow the evolutionary paths of the giant-supergiant group. This conclusion is supported by the fact that the pulsation energies in voluminous stars with tenuous interiors can become dissipated more easily than in denser stars with higher surface temperatures.

The last of the three groups of the young variables are the type I Cepheids. It has already been mentioned that the expected course of evolution of this class of stars on the H-R diagram is from right to left and downward toward the RR Lyrae variables. This evolutionary trend will be the outcome of increased densities which in turn will bring about corresponding reductions in the rates of pulsations associated with reduced volumes, higher temperatures and lower luminosities. The prerequisite to this evolutionary course is that these stars will be able to replenish their atmospheric gases. As the gravitational attractions between the stars in the Galaxy increase and the distances become shortened, most of them will be able to accumulate additional atmospheric material and keep on this evolutionary course. But some may not come across new supplies and the concentrations of their atmospheric atoms will become depleted slowly; the surface temperatures of these stars will drop, and their luminosities will become reduced. Consequently, they will move slightly to the right and downward on the H-R diagram, to produce the type II Cepheids. In other words, the aging of the type I Cepheids will cause them to split into two groups; most of them will become converted to RR Lyrae variables and the rest (approximately 2%)* will evolve into type II Cepheids.

---

*The ratio between the known type II and type I Cepheids is about 8%. The type II group comprises older stars. The total population of aged stars in the Galaxy is presumably about four times as much as the young ones. After correcting for these relative population densities, a value of approximately 2% is obtained.

At first glance, the relative abundance values of these stars do not seem to be of much help in confirming the suggested evolution of most of the type I Cepheids to RR Lyrae variables. But if we look at them more carefully, we find the values do correlate quite closely. The percentages in Table I have been calculated on the basis of the number of observed variables and not on relationship to the population densities of stars in the regions. The RR Lyrae pulsating stars are found among globular clusters, primary arms and perhaps the nucleus whose star population outnumbers that of the secondary arms by a great margin. Now, if the calculations are made on the basis of the ratio of the observed RR Lyrae to the total number of the older stars and then compared with the ratio of the observed type I Cepheids to the total number of the younger ones in the spiral arms, it will be found that the relative abundance of type I Cepheids in the spiral arms is in line with that of RR Lyrae variables in the rest of the Galaxy.

VII

THE EVOLUTIONS OF PLANETARY NEBULAE, WHITE DWARFS, WOLF-RAYET STARS AND ERUPTIVE STARS

Although the group of stars to be discussed in this chapter are not in the mainstream of stellar evolution, they do have a place in the scheme of the galactic evolution. They may be regarded as deviates from the normal stellar evolutionary course—that is, the stars in these groups are not following the paths taken by the majority. Besides being spectacular in nature, the significance of these stars lies in showing that no matter what route is taken by the minor members of an evolutionary system, all paths eventually lead to the same destination. In their case, they will most probably end by becoming converted to either cold, solid masses or dust particles. Their final fates, however, will be their integration into the atmospheres of massive stars and their conversion to energy, which is then radiated into space.

The origins of the members of this unusual group may be traced to binaries developed in the course of evolution of the Milky Way. As will be seen in Chapter IX, the secondary arms of all spiral galaxies are in the process of continual folding, the pulling in of the arms toward the nucleus being brought about by the gradual increases in gravitational attractions among the stars of the arms and the central structure; the interstellar distances become shortened and some stars become attracted to others. The result is the formation of a variety of binaries with diverse characteristics.

Since stars are made of either +*matter* or *antimatter*, the stars of the binaries could pair in one of the following three combinations: (1) a +*matter* star with a +*matter* star forming

155

a homogeneous +*matter* binary; (2) an *antimatter* star with an *antimatter* star producing a homogeneous *antimatter* binary; and (3) a +*matter* star and an *antimatter* star resulting in a heterogeneous binary. It is believed that, under specific conditions, planetary nebulae, white dwarfs, Wolf-Rayet stars and eruptive stars evolve from some of these binaries.

In general, four major factors have played parts in determining the present distances between the stars of the binaries: (1) the rate of increase in the gravitational attraction between stars (function of age); (2) the velocities of approach, a factor affected by the respective masses of the two stars, by their positions with respect to each other and by the influences of other members of the Galaxy; (3) the repulsion caused by radiation from both stars, effective especially at short distances; and (4) depending on whether they are homogeneous or heterogeneous binaries—the electrostatic repulsion or attraction originating from the ionized atoms whose lightweight electrons or positrons have escaped through the radiation zones. The attraction or repulsion by the electrostatic forces becomes a significant factor among the very short-distance binaries: if the components are homogeneous, the two stars cannot get too close to each other as the combined repulsion by the radiation energies and electrostatic forces effectively counteract the attractive gravitational forces; on the other hand, if the components are heterogeneous, the electrostatic attraction at short distances supplements the gravitational forces and pulls the stars very close to each other. This produces binaries in which the separation of the two stars is caused principally by the energy produced between the two antagonistic atmospheres.

The theories in this book indicate that planetary nebulae originate from some of the homogeneous short-distance binaries and that white dwarfs, Wolf-Rayet stars and eruptive stars evolve from specific types of heterogeneous short-distance binaries.

Planetary Nebulae

The evolution of planetary nebulae may be followed through a series of stages occurring among certain members of either

+*matter* or *antimatter* homogeneous short-distance binaries. Here, we shall describe the evolutionary progress of the homogeneous binaries, with both stars being composed of +*matter* central bodies and *antimatter* atmospheres. This is with the understanding that the same phenomenon can take place with binaries composed of stars with *antimatter* central bodies and +*matter* atmospheres.

The binary U Cephei seems to be in the initial stage of the evolution from a homogeneous binary to a planetary nebula. This double star is composed of a massive B8 star and a G-type subgiant. The mean distance between the surfaces of the two is about 4,500,000 kilometers, which is approximately the same as the diameter of the hotter star. A significant aspect of this binary is that the B8 star rotates much faster than the synchronization of the orbital and rotational periods of the two stars predicates. The discrepancy between the calculated and the observed rates of rotation suggests a continuous flow of gases from the subgiant to the primary; this conclusion is substantiated by spectrographic studies of lights from these stars. The probabilities are that the more massive and denser star is attracting and acquiring the atmospheric gases of its companion. There does not seem to be any obstacle in the way of this flow; consequently we should expect the transfer of gases to continue until the atmosphere of the red star becomes practically depleted and its radiation zone becomes ineffective in confining the contents.

When the atmosphere of the subgiant becomes almost exhausted, the contents will break through a weak spot and begin to leak out. Then, as the ruptured star orbits around the primary, its contents will become dispersed along its path to form a ring. Now, the material in the ring is composed of dense atoms which then become mixed with remnants of its atmospheric gases as well as some of the antagonistic gases from the star's atmosphere. As will be seen in the next chapter, these are the required conditions for the formation of luminescence in an interstellar cloud, and consequently, the ring will turn into a bright nebula.

The next stage of the evolution consists of the establishment of a state of equilibrium between the central star, its atmosphere and the ring. If we assume the central star to be composed of

+*matter*, the atmosphere will consist principally of *antiH*, and the nebula will be a mixture of +*matter* (the major component) and *antimatter* gases and dust particles. The ring will be highly rarefied on its outer edge and expanding outward, most dense in the middle and somewhat tenuous at the area of contact between the inner surface of the ring and the outer region of the *antimatter* atmosphere, where a weak radiation zone will be keeping the two layers apart. This type of structure is subject to a number of localized +*matter-antimatter reactions* at the border between the nebula and the atmosphere—which usually result in light flare-type eruptions. But whereas the solar flares are pulled back by the gravitational attraction of the Sun, in a planetary nebula the gravitational attraction from the central star complements the thrust and produces radial intrusions from the nebula toward the center. These spike-like structures may be seen clearly in the photograph of the planetary nebula in Aquarius (Figure 17).

The radiation pressure between the ring and the atmosphere of the central star forces the atmospheric gases in the middle to become more compact; this in turn increases the rates of reactions on the surface of the star and results in higher temperatures. Such hot stars are destined to use up their limited fuel supplies at fast rates. The depletion of the atmospheric gases is accompanied by a slow inward expansion of the nebula (Figure 18). When the star's original atmospheric gases become completely exhausted, then the central star begins to cool. In most cases the cold star becomes gradually covered with the bright nebula and lost from sight, but occasionally one of them may become observable by producing a dark comet-shaped pattern within the nebula (Figure 19). It has already been mentioned that bright nebulae are mixtures of antagonistic atoms and dust particles, with the luminescence coming from the energy produced by the reactions between the antagonistic atoms. The major component of the mixture in a planetary nebula is made of the same type of matter as the cooled star, and as the star moves relative to the cloud it reacts with the minor antagonistic component and removes the energy producing sources along its path, leaving a dark area behind. In the Figure 19 photograph, the black sphere is the cooled star

Figure 17. THE HELICAL PLANETARY NEBULA IN AQUARIUS *(Hale Observatories Photograph).* The spike-like intrusions are radially directed toward the central star.

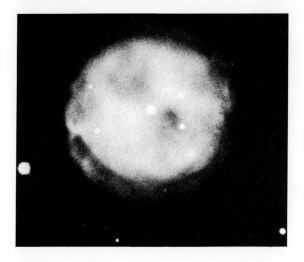

Figure 18. THE LAST STAGES IN THE EVOLUTION OF A PLANETARY NEBULA—THE OWL NEBULA. *(Hale Observatories Photograph).* Here the central star's original atmosphere is nearly exhausted and the bright nebula is in the process of closing in.

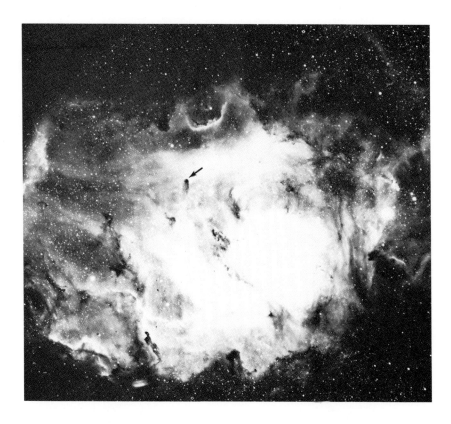

Figure 19. THE BRIGHT NEBULA MESSIER 8 (*Lick Observatory Photograph*). The arrow points to a cold star moving within the bright cloud. The dark comet-shaped tail is probably the result of the removal of the minor antagonistic component in the nebula along its path.

and the dark tail behind it the area cleared of the antagonistic atoms. This phenomenon, of course, is not restricted to the cooled stars of planetary nebulae; it can also occur when a relatively large cold, dense star becomes trapped inside a bright nebula (the major component of a nebula should be composed of dust particles of the same type of matter as the cold star).

White Dwarfs

These stars differ from the other members of the H-R diagram by their properties of high surface temperatures, low luminosities, very small volumes and extremely high densities. Their spectra are also unusual: a number of them have continuous spectra, which according to our theories is an indication of compact atmospheres composed of dense atoms. Broadened hydrogen lines have been observed from a fair number; in some the hydrogen lines are about the same as or somewhat broader than those found in the main sequence spectral A stars. And, in a few, in addition to fairly narrow hydrogen lines, there are lines of iron and other metals. The last two types of spectra point to the presence of normal atmospheric gases above the immediate dense atmospheres. The low luminosities of white dwarfs make them difficult subjects for observation; hence all our information comes from the nearby stars. But judging from their relative abundance in the neighborhood of the Sun, they seem to be rather common in the Galaxy, or at least in its spiral arms.

A very important and neglected phenomenon associated with white dwarfs is their occupation of an isolated island on the H-R diagram, a clear indication of their not being a part of the major evolutionary trend of stars. If they were, then with all the stars in the Galaxy, we should have some evidence of links to the main sequence. The absence of a connection with the general evolutionary trend points to a sudden separation from the mainstream of stellar evolution. This type of isolation is also seen in novae: soon after their eruptions they also break away from the main evolutionary channel and, similar to the white dwarfs, move to the left side of the main sequence.

The following is an analysis of the important and pertinent properties of the white dwarfs:

1. Since these stars do not have any links with the main sequence, either they began their lives as very dense, hot stars, or in their younger days they were members of the main sequence and because of some specific conditions suddenly ended up in their present positions on the diagram. Their low luminosities point to the second possibility; all the hot stars of the secondary arms have much larger masses and higher luminosities than the white dwarfs and it seems unreasonable to believe that some have deviated so radically from the pattern.

2. Most probably the extreme denseness and high temperatures of these stars are associated with the events that caused their sudden exit from the main sequence.

3. According to the theories in this book, in order to attain the very dense interiors and maintain the relatively high temperatures, the white dwarfs should possess atmospheres composed of dense atoms; these could come only from the central bodies of other older stars.

4. Their possession of dense atmospheres suggests they originated from short-distance heterogeneous binaries containing two old, red dwarfs.

Precursors of White Dwarfs

One group of binaries, the W Ursae Majoris, seems to possess the requirements for evolving into white dwarfs. Following are the reasons for this belief:

1. The masses of the components in these binaries are comparable to and usually smaller than the mass of the Sun. Their conversion to white dwarfs results in the destruction of part of their masses. When this loss is taken into account, the sum of the masses of the two red dwarfs falls within the range of a white dwarf's mass.

2. The members of this group have very short periods of revolution, usually between 5 to 12 hours; this indicates their extreme proximity and the possibility of physical interactions between the atmospheres of the components.

3. It has been suggested from the studies of the spectra of this group that a belt of luminous gases exists between the two components.

4. The spectrographic observations point to both components being red dwarfs.

Formation of White Dwarfs

The mechanism for the transformation of a W Ursae Majoris binary to a white dwarf would entail the unification of the two red dwarfs. The first stage would consist of the total exhaustion of the atmosphere of the smaller component by its continual interaction with the atmosphere of the larger one, the two being antagonistic to each other. In the next stage the released interior gases from the secondary are attracted to the primary and surround it. Since the original atmosphere of the primary and the contents of the secondary are of the same type of matter, the new atmosphere does not encounter any opposition. As the dense gases from the body of the secondary turn into the atmosphere of the primary, the high-density atoms of the new gases gradually replace the original atmospheric low-density atoms and push them to the higher levels in the form of a tenuous shell composed of normal lightweight atoms.

It is believed the displaced original atmospheric atoms are the sources of the varied spectral lines observed in the lights of different white dwarfs. Each of these dense stars has gone through evolutionary changes of its own and consequently the composition of the light atmospheric shell is not about the same for all of them; some do not even possess this cover of the light gases. The basic spectrum for all the white dwarfs is the continuous one coming from the luminescent, dense atmospheric material; the absorption lines indicate the positions of the original light gases with respect to the star.

Causes of High Densities

At the time of a white dwarf's formation, its interior atoms do not have the extremely high densities found in most of these stars. It takes long periods of intense energy production on the surface to bring up the values to such fantastic figures. In the beginning, both the atmosphere and the body are composed of atoms of medium density. Though these conditions cause the creation of a powerful radiation zone and bring about a considerable rise in temperature, they cannot suddenly change the densities of the interior atoms. However, at the time of transformation, the intensified reactions on the surface and the high densities of the atmospheric material do bring about some contractions of the volume and with them a proportionate reduction in luminosity, higher tempeatures and a corresponding increase in the density of the star. This by itself is not particularly significant. The most important factor that causes the raising of the densities is the very intense reactions on the surface carried on for a long period of time. The principle of this mechanism has already been explained.

It does not seem feasible to assume that the gain in the densities will continue during the entire life of these dense stars. Eventually the time comes, as it has probably done already in a number of white dwarfs, when the physical conditions prevent any further increases in the masses of the subatomic particles and the densities of the atoms. From then on there will not be any significant change in the densities of these stars, and in most cases they will continue with their monotonous existence until their atmosphere becomes exhausted. Then they become converted into extremely dense and relatively small solid masses which in the course of time will either become components of some of the eruptive stars or follow the path already described for the conversion of cold stars to the atmospheres of massive stars.

Statistical Considerations

The estimated number of white dwarfs is considerably higher than that of W Ursae Majoris stars, but this difference is caused by their life spans. As contrasted to white dwarfs, the W Ursae Majoris binaries have short lives. The longevity of white dwarfs has a cumulative effect. The binaries are in a temporary state between their existence as red dwarfs and the white dwarfs of the future. Hence they are less numerous than the stabilized dense stars which have become accumulated throughout the ages.

Wolf-Rayet Stars

These are high luminosity (absolute magnitude -4 to -8) stars with very high surface temperatures (40,000 to 100,000° K), emitting exceedingly broad emission lines of ionized helium and doubly and triply ionized nitrogen, oxygen or carbon. Nearly 200 of these W stars are known. They are about two to six times as large as the Sun and have approximately ten times the solar mass. The most peculiar characteristic of these stars is their continual ejection of very hot gaseous material into space, indicating internal instability.

In terms of the proposed theories, these properties indicate them to be relatively massive stars (about class B) that, instead of having normal atmospheres, are in possession of high-density atmospheres similar to those of the white dwarfs. Their instability then may be ascribed to the inability of the internal gases to remain in a stable condition when subjected to the very intense reactions on their surfaces. Eruptions break through the very dense and well-packed atmosphere. As a result, vast amounts of energy are produced and great quantities of mixed antagonistic gaseous material are ejected into space.

A star of B class can acquire a high-density atmosphere either through the accumulation of large quantities of very dense dust particles from the interstellar clouds, or through the acquisition of material from the central mass of another star. The first possibility seems to be inapplicable to the Wolf-Rayet stars; the gathering of large quantities of high-density material from galactic dust is a slow process and is subject to continual readjustments which prevent the formation of highly unstable conditions. The more plausible explanation is the acquirement of the contents of a neighboring star via the short-distance heleterogeneous binary route. If that is the case, then we should look for binaries that contain B stars as one of their components. The Beta Lyrae group of eclipsing variables seems to meet these conditions. Beta Lyrae itself, as judged from its relatively long

period of 12.9 days, is quite a distance away from being transformed into a W star, but another member of this group, V
Puppis, seems to have the qualifications to be the first of the
known binaries to become converted to a Wolf-Rayet star.

Observations indicate that V Puppis is composed of a B1
primary and probably a B3 secondary with a period of approximately 1.45 days. This is a comparatively short period, especially when one includes the huge masses of the components;
this indicates the extreme closeness of the stars and points to the
possibility of their conversion to a W star in the not-too-distant
future. The expected mechanism for the transformation of this
primary to a W star is about the same as the one described for
the white dwarfs: the two antagonistic atmospheres react with
each other until that of the smaller component is exhausted; the
released high density material from the interior of the secondary
is then attracted to the primary to become its new atmosphere.
The new acquisition brings about the intensification of the reactions on the surface and the characteristic instability that is the
mark of the W stars. It should be remembered that in the conversion of this type, large quantities of material are lost during
the adjustment period; therefore one should not expect the mass
of the resulting star to be equal to the total mass of the two
components.

It does not seem likely that the Wolf-Rayet stars can remain
in their hyperactive states for long periods of time. The very
high rate of energy production and the ejection of large quantities of matter into space is bound to lower the concentration of
the atmospheric atoms; this would in turn bring down the surface temperatures correspondingly. Therefore, the probable
evolutionary course for these stars will be their slow reconversion to stable stars. Once the excessive activities are over, the W
stars will most probably move toward the blue giants on the
main sequence and then follow more or less the same evolutionary paths as other members of this group.

Eruptive Stars

The accumulated evidence points to the great possibility that
the short-distance heterogeneous binaries are the precursors of

the eruptive stars, with the nature of the stars determining the the type of eruption. Here we shall concentrate on explaining the mechanisms of eruptions. The discussions will be limited to novae and the two types of supernovae. There is not enough information about other members of this group to give guidance as to their origins and the causes of their eruptions.

Novae

In general, each nova originates from a star of low magnitude. The star brightens up at a rapid rate and within a few days reaches its maximum brightness; after that it begins to fade, taking many years to reach its original luminosity. For example, Nova Aquilae was an insignificant star before June 7, 1918, when it brightened up to a magnitude of +6 in one night, reached a magnitude of +1 by the next night and brightened to a visual magnitude of -1.4 (absolute magnitude of approximately -9) by the third night. But within the next 24 hours its luminosity dropped to 0, and after that it gradually faded until 1924, when it reached its original brightness. Spectrographic observations showed that the explosion had resulted in a luminous shell which was expanding at the radial velocity of about 800 kilometers per second. The rise in luminosity in the first three nights was caused principally by the very high temperatures combined with the expansion of the luminous gases. The decline came with further expansion, which brought about the rarefication of gases and the lowering of temperatures. (It should be mentioned that the word "shell" is a questionable term. Later observations on other novae have shown that the gases do not expand uniformly in the form of a shell, but rather in two diametrically opposite directions.)

Most of the data about the nature of the quiescent stars of this group have been gathered from observations of the old Nova Aquilae (1918) and Nova Herculis (1934), with the latter providing the major part of information that relates novae to binaries. The old Nova Herculis is a binary composed of a white dwarf primary of 0.12 times the solar mass and a red star secondary of 0.20 times the mass of the Sun. A stream of high-temperature hydrogen gas is apparently flowing from the sec-

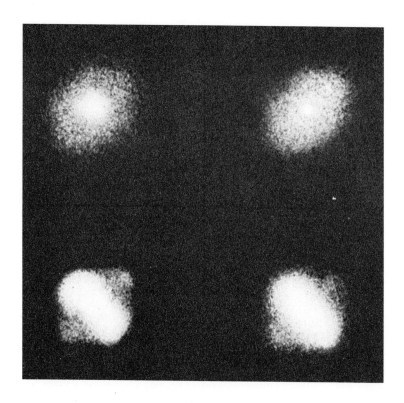

Figure 20. NOVA HERCULIS (1934) AS OBSERVED IN 1942 (*Hale Observatories Photograph*). Top left: In blue violet light. Top right: In the green N_1 and N_2 lines of doubly ionized oxygen. Bottom left: In the red light of ionized nitrogen (taken in July, 1942). Bottom right: In the red light of ionized nitrogen (taken in August, 1942).

ondary to the primary. On the basis of these observations, suggestions have been made that the gas is being gathered around the white dwarf in the form of a shell (or more probably a disc) and that this accumulation is somehow responsible for the explosion.

The general properties of this binary, which has already had one recorded explosion and which will probably have at least one more in the future, fit into the picture of a heterogeneous short-distance binary composed of a white dwarf with a dense atmosphere and a red dwarf with a light atmosphere. Even though the white dwarf has a smaller mass than its companion, its very dense and compact nature is acting as a point of attraction for acquiring the hot antagonistic gases from the upper atmosphere of its companion. The accumulated gases form a wide ring of lightweight antagonistic gases around the white dwarf over the densely packed atmosphere. This ring is probably concentrated on the equator while the poles attract almost nothing. As the gases are accumulated over the white dwarf's dense gases, +*matter-antimatter activities* increase between the two layers. Eventually, a triggering mechanism (such as a flare or a prominence) starts the explosion, but this will occur only after appreciable amounts of atmospheric gases have moved from the secondary to the primary. The explosion causes the intermingling of the antagonistic gases and the expansion of the mixture at a very fast rate. As the gases expand, collisions of a large number of high-velocity antagonistic particles bring about their annihilation and the generation of great amounts of energy. Eventually the expansion reaches the diffuse state where the chances of collision between the antagonistic particles become reduced and the luminosity begins to turn downward. However, the expanding gases continue to generate energy for a number of years.

The released energies do not seem to be strong enough to overcome the gravitational and electrostatic attractions between the two components of a recurrent nova and cause their permanent separation. The force of the explosion probably makes them move apart somewhat, but as the activities subside, the forces of attraction between the two gradually bring them back together to start the next cycle.

Type II Supernovae

With the exception of the intensity and the grandeur of the explosion, the properties of Type II supernovae, especially the spectroscopic characteristics, seem to be very similar to those of novae. However, the luminosities and mass losses are quite different: whereas the absolute magnitudes for novae seldom exceed -8 and their mass losses are relatively small, Type II supernovae attain absolute magnitudes of as high as -14 and their main losses are between 1/100 to 1/10 of the total mass of the binary. These properties suggest the mechanism of the outburst to be fundamentally the same for both groups. If this be the case, then the differences may be traced to the size of the explosion: a nova usually has a number of blowoffs at intervals; with a Type II supernova, one grand and final outburst completes the act.

A Type II supernova most probably differs from a nova in the different composition of the atmosphere of its secondary and the manner of accumulation of the gases by the primary. In novae the gases are collected in the form of a wide ring; in Type II supernovae the gases spread over the surface of the primary in the form of a shell. This can happen if the density of the secondary is high enough to prevent its shape from becoming roughly elliptical under the effects of the gravitational attraction from the nearby dense primary. When the density of the secondary is not high, as is the case with a red dwarf in a nova binary, it becomes distorted in shape and the atmospheric gases are pulled in by the primary through the sharp point created on the surface of the secondary. The continuous accumulation of the thin stream of gas coming from this point produces a ring around the dense mass. However, if the secondary is dense and its central body not disfigured (as can happen with a young white dwarf), the gases are removed from the surface of the secondary and spread more or less evenly over the dense atmosphere of the older primary. The continuation of the accumulation will eventually lead to an explosion triggered by an eruption between the two layers. This results in the very rapid mix-up of the two relatively dense antagonistic gases; the final outcome is a huge explosion.

Type I Supernovae

Although no sharp lines can be drawn between Types I and II supernovae, the following provides a summary of differences between the two as outlined by Baade and Minkowski:

1. Type I supernovae are considerably brighter than Type II; the luminosities in the latter seldom reach an absolute magnitude of -14; the absolute magnitudes of Type I vary between -14 and -17.

2. Type I supernovae have sharper maxima than Type II and the decline that follows proceeds at faster rates; this indicates more powerful explosions in Type I, and faster rates of expansion of the resulting products.

3. The masses of the components of Type I binaries are as a whole smaller than in Type II. But the percentage of loss of mass after the explosion is greater than in Type I (about 1/10 to 9/10 against 1/100 to 1/10 for a Type II).

4. Type II supernovae produce broadened emission lines of hydrogen and several other ionized atoms; Type I supernovae are deficient in hydrogen and have overlapping bright lines that as yet have not been identified. These properties point to the probability that the components in Type I binaries are older and denser than the ones in Type II binaries.

A fundamental pattern emerges from the comparison of the composition of the binaries of different types of erupting stars: all are heterogeneous, very short-distance binaries with white dwarf primaries, and the magnitude of eruption is related to the degree of denseness of the secondary. Accordingly, we should expect the secondary in a Type I binary to be an extremely dense star—that is, the binary is probably composed of two heterogeneous old white dwarfs.

In a Type I binary, two very dense stars are thus being attracted to each other by extremely strong forces, but are kept apart by the energies produced at the area of contact between the two atmospheres. The reactions between the two antagonistic layers bring about the gradual depletion of both atmospheres, with the less massive atmosphere of the secondary being the first to become exhausted. But since the secondary is composed of a very dense mass, it solidifies before its atmosphere becomes completely depleted. After the removal of the protection of its atmosphere, the very dense and solid secondary is pulled into the primary. The atmosphere of the primary, being of the same type as the body of the secondary, does not resist the intrusion. In the encounter between the two antagonistic central bodies, the smaller and cooler solid mass of the secondary penetrates into the hot gaseous primary with the Leidenfrost layer protecting it for a short time. Once inside, it becomes gaseous on the surface; the gases expand and eventually break through the surrounding radiation zone and mix with the contents of the primary at an extremely rapid rate within a confined space. The resulting cataclysmic outburst produces a very fast expanding nebulosity composed of a mixture of antagonistic dust particles. The explosion may either pulverize the two dwarfs completely or blow their remnants apart within the newly created nebula.

Similar to any explosion, the disintegrated material in the nebula is not distributed evenly. Then, not only do the concentrations of the particles differ in various locations, but the ratios of +matter to antimatter particles vary greatly from region to region. Consequently, the expanding nebula develops streaks, patches and uneven luminous areas, with the degree of energy production depending on the ratio of +matter to antimatter dust particles in the region. These variations may be seen in the Crab nebula (Figure 21), which exploded about a thousand years ago. In addition to showing these characteristics of unevenness the nebula also emits polarized electromagnetic waves and radiates powerful pulses from an extremely dense mass in its central region.

Crab Nebula

The Crab nebula consists of a featureless amorphous nebulosity enmeshed within a network of luminous filaments. The substratum emits polarized light with no particular spectral characteristics and the filaments produce bright spectral lines, denoting intense rates of energy production. Since the two features are components of one structure and have been in-

Figure 21. THE CRAB NEBULA (*Hale Observatories Photograph*). Two stars are located in the central region of the nebula, one of which is a pulsar. Neither can be recognized in this photograph.

fluencing each other for a relatively long time, they have developed a certain degree of inter-relationship. Most probably the intensely powerful radiations within the filaments are creating conditions that are causing the production of polarized lights from the lighter particles in the amorphous stratum. This means that a cycle has developed where the energies from the filaments are directing the motions of streams of particles along certain routes in the nebula. While the particles are driven away from one of the filaments at great speeds, the next one causes them to change direction with the one after bringing about new alterations. As a result of about a thousand years of this type of activity, a cycle has been established in which the particles move along certain directed paths, while constantly changing speeds and directions. These are the conditions for the emission of polarized light. But since the movements and the changes of directions take place in three dimensions and follow complicated patterns, it is difficult to discern their true courses in the two-dimensional photographs.

The second interesting aspect of the Crab nebula is the presence of two small stars of low magnitude in its central region, with one of them emitting electromagnetic waves in pulses at intervals of 0.033 seconds. The proposed hypothesis regarding the Type I supernovae explosions accounts for the presence of these two extremely dense objects; they are the remnants of the two antagonistic white dwarfs that were involved in the explosion, with the pulsar being the rotating, disfigured secondary.

In the explosion of the Crab supernova, the entry of a cooled secondary into a hot primary caused two important changes in the secondary: first, a significant portion of its mass was blown away, particularly on the side that was facing the explosion (the primary, having been in the gaseous state, repaired its spherical shape); and second, two types of motions appeared in the solid mass, a linear one away from the primary and a rotary one (the rate varies with the conditions of the explosion) around the center of gravity of the deformed mass. The energies from the explosion did not have a long enough time to heat the solid mass of the secondary to a gaseous state. The two masses, on their part, could not move away from the center of the explosion with the velocity of the expanding nebulosity. Two factors were responsible for this slower rate of movement: first, the gravitational attraction between the two very dense masses was still effective; and second, the continual encounter with numerous antagonistic particles in front of them produced sufficient energy to reduce the rates of their linear motions. The effects of these two factors may be seen in these two stars; by now neither has any significant linear motion relative to the expanding nebulosity. However, the angular velocity of the secondary was not particularly affected by these hindrances. Although the rate of its rotation has been reduced, it still possess a fair portion of its original spin. This rotation is believed to be the cause of emission of electromagnetic waves in pulses.

The Crab Nebula Pulsar

The pulsar in the Crab nebula is rotating within an expanding luminous nebula composed of a mixture of +matter and antimatter dust particles darting back and forth and colliding with the rotating mass when it is in their paths (the details of the structure of luminous nebulae are discussed in the next chapter). If a colliding particle is of the same type of matter as the mass it strikes, no special effects are produced; but if it is antagonistic, it is annihilated and an equivalent amount of energy is generated and radiated into space. Since the mass is rotating at a fast rate, the contour becomes an important factor in determining the rate of energy production on the different areas on its surface. If the mass is spherical (as is the case with the primary), the collisions are about equal on all sides and there is a continuous and uninterrupted radiation in all directions. However, if it is deformed, such as having a ridge which is higher than the average distance from the center, then as it rotates at the rapid rate (the present rate—one rotation every 0.033 seconds), the elevated part sweeps the particles away, creates a vacuum behind it and significantly reduces the rates of collision of the antagonistic dust particles on the lower part of its body. Since the entire width of the ridge is open to collisions, the elevated area becomes the principal source of electromagnetic waves. Accordingly, the pulsar in the Crab nebula is radiating its signals from the elevated section (relative to the center of rotation) on its surface, and the duration of the pulse represents the time the raised part is taking to pass over our line of sight.

It should be remembered that the pulsar is rotating within an expanding nebula with varying conditions. At times it may encounter large amounts of antagonistic atoms or it may be covered by a dust cloud with low antagonistic particles contents; it may come across currents of particles that would impede or accelerate its rate of rotation, and the movements of particles may interfere with the nature of its radiation (e.g., polarize it). Consequently, we should not expect continuous uniformity in the pulse frequency or the type of radiation we receive. However, the indications are that the mechanism of energy production of the pulsar tends to slow down its rate of rotation. The energy generated on the surface of the ridge and particularly at its very edge, where most of the particles strike, opposes the rotation of the mass. For this reason, the rate of rotation of the pulsar is on the decrease and the lengths of the periods between the pulses are becoming longer with time.

If the rotation of the pulsar in the Crab nebula has been slowing down, then at the time of its ejection it ought to have been rotating at a much faster rate than its present one. This raises the question of the possibility of its disintegration under the influence of the immense centrifugal forces at that time. But we are not in a position to pass judgement on such a possibility, for we do not have any knowledge of the structures of the atoms and the natures of the interatomic forces in these extremely dense masses; therefore, no predictions can be made regarding the angular velocities at which disintegration would take place. So far, the only hope of gaining an insight into the structures of these atoms lies in the successful capture and the study of the very dense cosmic particles.

VIII

THE EVOLUTION OF NEBULAE

The principal difference between nebulae and interstellar dust lies in the concentration of dust grains per unit volume of space. When the number is low, as is the case with interstellar dust, the light from distant stars passes through with only minor alterations. However, when there are high concentrations of dust grains, as is the case with nebulae, the passage of starlight is often obstructed in varying degrees and sometimes completely. As a result, a cloud-shaped structure is frequently observed against a star-studded background.

Similar to atoms in gases, the dust particles in space are under the influence of two principal factors, the repulsion effects of radiation energies and the attractive forces of gravitation. If the dust particles are composed of relatively low-density atoms and are thrown into space at high temperatures and velocities (such as in novae and Wolf-Rayet stars), the gravitational forces between the grains cannot hold the particles together, and the grains separate and spread in space to become interstellar dust. But if the dust originates from high-density stars through explosions (as in supernovae), or from the interior of stars where the radiation energy per grain is not great enough to overcome their gravitational attraction (as in planetary nebulae), then the particles hold together to form a nebula.

Sources of Nebular Material

The evolution of planetary and supernova nebulae has already been described. Nebulae can also originate from red

177

giants and supergiants. In chapter VI the path of evolution of these stars on the H-R diagram was indicated to be from right to left and downward, but this is contingent to their continual accumulation of new atmospheric material. However, there are always some that do not get the chance to replenish their consumed atmospheric gases. Then, similar to the red subgiant companion in the binary of a planetary nebula, their gaseous contents break through weak areas, mix up with what is left of their atmospheric gases and turn into nebulae. The densities of dust particles in different nebulae originating from red giants and supergiants do not need to be approximately the same. Some of these massive stars might even have been of G or F classes; then, as their atmospheres became depleted, they moved to the right on the H-R diagram. These would produce dust clouds with higher-density components than the original M stars.

Considerable amounts of hydrogen, helium and other elements of higher atomic weights (or their *antimatter* equivalents) can become trapped inside nebulae evolving from red giants and supergiants. The light atoms may come from their atmospheres or be pulled in from interstellar gases by the gravitational pull of the nebula; once entrapped, they are retained within the cloud by the strong gravitational forces from the dust grains. However, the presence of light gases in a nebula does not necessarily indicate their giant-supergiant origin, as the interstellar gases can also be accumulated by any nebula containing high-density dust grains. Such nebulae exhibit the properties of both gas and dust. For example, they interfere with the passage of light from stars behind the clouds, and at the same time radiate 21 cm. neutral hydrogen lines into space.

Generation of Electromagnetic Waves in Nebulae

It is evident that the concepts in this book do not rely entirely on the principle of photoionization to explain the generation of electromagnetic waves by nebulae. The following list provides examples where the theory of photoionization does not provide convincing explanations:

1. Although the spectacular bright nebulae, such as the Orion

nebula, are close to hot stars, there are many bright nebulae which either do not have a companion B or O star, or the surface temperature of the nearest star is too low to produce the quantities of ultraviolet radiation needed to bring about the photoionization of the whole cloud.

2. If a nearby star is responsible for the production of light in a bright nebula, then all the bright nebulae should have the same general pattern of distribution of light as the Orion nebula (Figure 22)—that is, being brightest on the side facing the star and becoming darker with distance from the source of radiation. This type of pattern is seen only in a limited number; most bright nebulae have either no definite pattern of distribution of bright regions relative to a nearby star or generate light almost uniformly from their entire surfaces.

3. If the luminescence of bright nebulae is due to the recapture of electrons by ionized atoms, then no dark clouds should be present near the bright nebulae or their stars. In many nebulae, such as the bright and dark nebulosities near Gamma Cygni (Figure 23), North America and Pelican nebulae (Figure 24), and dark nebulosity in Monoceros (Figure 25), patches of dark nebulae are intermingled with bright ones. Naturally, some are in front of the bright nebulae and may be claimed as being too far away from the source of radiation to receive the energy needed for photoionization; but does this explanation hold for all the dark nebulae neighboring the bright ones?

4. We receive helium emission lines from a number of bright nebulae, indicating intense energy production of the order of hot stars and the solar chromosphere. The solar helium emission lines are produced only in one region. The Sun is not a hot star, but it does produce high-energy ultraviolet radiation, the intensity of which at close range is just as strong or stronger than the intensity of the same type of radiation from hot stars at the astronomical distances that separate them from their respective nebulae. Yet, with the exception of the chromosphere region, the radiation from the Sun does not produce any emission lines from the helium atoms of its own atmosphere, not even in the regions where the concentration of atoms is the same as those in the bright nebulae.

5. The brightest stars of the Pleiades cluster are surrounded

Figure 22. THE ORION NEBULA (*Hale Observatories Photograph*).

Figure 23. BRIGHT AND DARK NEBULOSITIES NEAR GAMMA CYGNI (*Hale Observatories Photograph*).

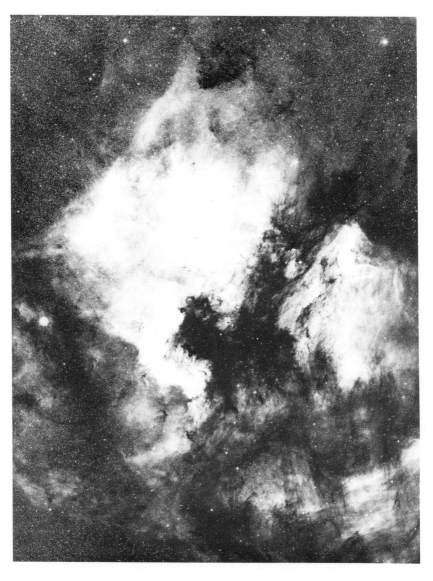

Figure 24. THE NORTH AMERICA AND PELICAN NEBULAE (*Hale Observatories Photograph*).

Figure 25. NEBULOSITY IN MONOCEROS (*Hale Observatories Photograph*).

by nebulosities that have the distinct property of producing absorption spectra. These nebulae are as close to the stars as any nebula can get. They receive very intense ultraviolet rays at very short distances; yet instead of emission lines they produce absorption lines. Now, if the ultraviolet lights of a hot star at close range cannot produce emission lines, what then causes the production of emission lines in some of the nebulae that are far from stars?

6. The photoionization theory cannot give the reason for the generation of continuous spectra, the major portion of emissions from many bright nebulae.

An alternative to the theory of photoionization is the concept of generation of energy inside the nebula by slow +*matter-antimatter reactions*. The principle of the hypothesis is simple: all nebulae are the by-products of evolution of some of the stars; as a result they are composed primarily of a mixture of antagonistic high-density dust grains. The degree of intensity of energy generation then may be explained in terms of the ratio of the two types of matter among the dust grains and the measure of compactness of these large particles within the nebula.

Since the mode of distribution of +*matter* and *antimatter* material in a nebula is not the same as in stars, the mechanism of annihilation of matter within these clouds is also different. Whereas in stars the two antagonistic layers are segregated, in nebulae the two types of dust grains are mixed. In stars, total annihilation of atoms entering the antagonistic zones causes the generation of intense energies; in nebulae there are no separated zones for different types of matter; the dust grains usually approach each other at non-relativistic velocities and are free to move away. Consequently, encounters between antagonistic particles usually result in mutual repulsions. The energy needed for the change of direction of motion comes from the annihilation of very small portions of masses of the encountering grains. The annihilation of a few atoms from each grain produces more than the needed energy; part of the energy is used up for repulsion and the rest is emitted as electromagnetic waves, part of it in the form of continuous spectra.

Seldom, if ever, is the ratio between the two antagonistic dust grains in a nebula even; there is always a major and a minor component. The nearer the ratio of the two is to unity, the brighter is the nebula. In each encounter both particles lose infinitesimal parts of there masses, but since the grains of the major component have smaller chances of encountering the minor component, the loss of their masses is on the average less than that of their opposite type. Consequently, the grains of the minor components become smaller in size at faster rates and their rates of disappearance are higher than those of the major one. With each loss, that much less energy is produced; therefore, as a nebula grows older it becomes darker. However, the rate of the loss of brightness due to aging is not the same in all nebulae; the ones with the ratio of their components nearest to unity are the slowest in depleting their minor components; hence the reduction of their brightness proceeds at slower paces. By the same token, the dimmer a nebula is at the time of its origin, the faster is the rate of the loss of its energy production.

The Nebular Forms

The expansion of many bright nebulae may be attributed to the internal production of large amounts of energy. The young ones, such as the Crab nebula, expand at comparatively rapid rates, but as they grow older and the rate of energy production becomes reduced, expansion subsides gradually. Eventually a balance becomes established between the gravitational forces from the dust grains and the rate of energy production. Thereafter, little expansion takes place; this equilibrum is maintained until the diminution of energy production reverses the trend. Then the nebula begins to contract at a very slow rate. In addition to internal causes, external effects can also alter the structural form of a nebula. These influences may be divided into three categories: (1) a powerful external source of radiation causes contraction of volume and increases the rate of energy production on the side of the nebula facing the star; (2) the encounter of two nebulae antagonstic to each other limits the expansion at their borders; and (3) a powerful source of gravitational forces at a close range reduces the rate of expansion. The best example of a bright nebula affected by radiation from a nearby hot star is the Orion nebula (Figure 21), in which the brightest region is closest to the star. It is evident that the powerful radiation from the hot star is producing the effect. This may be ascribed either to the photoionization effect or to the compacting effect brought about by radiation. Probably it is a combination of the two. In the compacting effect, the radiation pressure drives the particles away from the star and increases the chances of encounters between the antagonistic grains. It is interesting to note that in the Orion nebula most of the brightest regions facing the star have abrupt edges; this effect contrasts with the dispersion of the nebula in the regions farthest away from the source of radiation. Thus, one may observe that the greater the distance is from the star, the more dispersed is the cloud. The nebula is obviously expanding in one direction, away from the star.

The encounter between the two antagonistic nebulae occurs when the major components of the two nebulae are composed of opposite types of matter. It is characterized by the intermingling of many antagonistic particles at the border between the two nebulae and the generation of more than usual amounts of energy. This condition can cause the formation of a bright rim which will become quite pronounced if a dark nebula silhouettes against a bright nebula. The same effect may be produced if an intense source of light is behind a dark cloud. We do not have any reliable means for differentiating between the two types. The Horsehead nebula (Figure 26) seems to be of the second type; it is very dark in front and the bright rim is not pronounced. The M16 nebula in Serpens (Figure 27) is probably showing the encounter of antagonistic nebulae. The brightness of the rims seems to be more intense than if produced by reflected light, and there are distinct differences between two groups of small dark nebulae against the same background. The nebulae in one group have very bright rims and the dark clouds in the second group have no rims at all.

There are a number of examples where powerful gravitational forces from nearby sources influence the structural forms of bright nebulae. The planetary nebulae and the reflective ones associated with the stars of the Pleiades cluster are the prime examples of this type of effect. We have already given an example of how gravitational forces from a central star can for

Figure 26. THE HORSEHEAD NEBULA IN ORION (*Hale Observatories Photograph*).

Figure 27. THE M16 NEBULA IN SERPENS (*Lick Observatory Photograph*). The antagonistic bright nebula is preventing the expansion of the dark nebula and producing bright rims at the borders between the two types of nebulae. Also, the comparison of the three small, dark nebulae on the upper right corner with a group of small, dark nebulae on the upper left side indicates their relative compositions: the pronounced bright rims on the former group and no bright rims on the latter group suggest the possibility of the former being composed of the type of matter opposite to that of the dark clouds on the left and the major component of the bright nebula in the background. Behind the small dark clouds, the nebulosity has about the same amount of brightness.

some time control the shape and appreciably reduce the rate of expansion of a bright-ring planetary nebula. The example of reflective nebulae may be found among the stars of the Pleiades cluster; here these bright stars have created the unusual phenomenon of dividing one bright nebula among a number of massive stars through the effects of their gravitational attraction (Figure 28).

The original nebula most probably was the product of a supernova explosion; as the bright nebula expanded it was divided among the members of the Pleiades cluster. Had the major component of the nebula been of the same type as the atmospheres of the stars, the broken nebula would have been incorporated into the atmospheres. But since this did not happen, most probably the major component of the nebula was antagonistic to the atmospheres of the stars. This phenomenon of a bright nebula covering a hot star is in principle very similar to that of planetary nebulae; the difference between the two lies in the mode of their origin. In a planetary nebula, such as the M57 in Lyra, the source was a secondary in a homogeneous binary which released its contents to form a smooth ring as it circled around the primary. In the Pleiades cluster, all the nebulae are the remnants of a supernova which most probably exploded inside the cluster, with each star attracting and holding on to a portion of the streak-laden expanding nebula. In all cases a star is separated from the nebula by an atmosphere which is antagonistic to both. Consequently, unless they accumulate new material from the Galaxy, the atmospheres of the stars of the Pleiades cluster are going to be exhausted at fairly rapid rates.

The destiny of all nebulae, irrespective of their origins, is always the same: as the Galaxy ages and the distances between the stars are reduced, eventually all nebulae will come within the reaches of the gravitational forces of the more massive stars and become sources of material for the replenishment of their depleted atmospheres. Only one condition has to be met for the incorporation: the major component of the nebula should be of the same type as the atmosphere of the star. The length of the

Figure 28. SOME OF THE STARS OF THE PLEIADES CLUSTER COVERED WITH STREAKED NEBULOSITIES (*Official U.S. Naval Observatory Photograph*).

life of a nebula is therefore dependent on its distance from the nearest massive star; the shorter the distance to the right type of star, the sooner it becomes converted into atmospheric material. This explains why nebulae are abundant in the secondary arms, are infrequently observed in the primary arms and are very rarely seen in the nuclei. In the last two regions, the closely packed older, massive and dense stars, with their partially exhausted atmospheres, have great need for nebular material; whenever the opportunity arises they literally devour the approaching nebula. In the nuclei of galaxies, the competition for the removal of the debris of an explosion is particularly fierce. Unless the outburst is so great as to propel the contents beyond the reaches of the remaining members, the disintegrated materials of the old companions are divided up at great speeds. Consequently, the probability of a supernova explosion turning into a nebula is very remote inside the nucleus. The globular clusters also come under the same rules. Because the conditions in these clusters are somewhat similar to those in the nucleus and primary arms, there is little possibility of their containing nebulae.

IX

THE EVOLUTION OF GALAXIES

In this chapter, attempts will be made to inter-relate the evolutions of different classes of galaxies with the proposed hypotheses in this book. As usual, the emphasis will be placed on the interwoven relationships of evolutionary systems. In this approach the evolution of galaxies is considered to be the outgrowth of the evolution of stars of which they are composed. In order to lay the foundation for this discussion, we shall first describe the structural parts of a mature galaxy; this will be followed by a review of Hubble's system of classification. The evolutionary stages of different types of galaxies will then be discussed in terms of the principles outlined in previous chapters.

General Anatomy of Galaxies

A very obvious yet very significant fact about galaxies is that, with the exception of the irregular, and some of the "peculiar" ones, almost all of them have densely packed central cores surrounded by stars whose population densities decrease with distance from the center. In elliptical galaxies the reduction in stellar concentration proceeds smoothly; in spiral ones specific patterns have developed in the distribution of stars and dust clouds.

A mature spiral galaxy is composed of five principal parts. The most important three have been marked on NGC 3031 in Figure 10. The center and the hub is the nucleus. It is the brightest section of the unit but its borders cannot be defined, especially on face-on photographs. It consists of a large number of massive old stars packed in the smallest possible space. The densities of these stars, particularly in the core, are probably as high or higher than those of old white dwarfs; the only reason for the nucleus not collapsing is the intense radiation generated by its members. The nucleus is attached inconspicuously to what Hubble has called "incipient spiral arms which are usually smooth in texture and are tightly wound in nearly circular patterns about an amorphous central region." With the exception of the Sc group (see the classification which follows) whose spiral arms arise directly from the nucleus, the bright central core is almost always attached to these inner arms. We refer to these arms as primary arms. They can be seen clearly in such young galaxies as NGC 5273 and NGC 7743 (Figure 40). In the more mature galaxies, the primary arms are situated between the nucleus and the spiral arms. The third part of a spiral galaxy is composed of the conspicuous spiral arms which in addition to the relatively young stars contain large quantities of dust clouds; we call these "secondary arms," the term signifying their late arrival in a galaxy's development. The fourth and fifth parts of many spiral galaxies consist of globular clusters and stars with independent orbits; these do not follow the motions of galactic planes.

Classification of Galaxies

The following excerpts from the *Hubble Atlas of Galaxies**, compiled by Allan Sandage, describe the principal properties of most of the observed galaxies. The page numbers in the copied text refer to pages in the *Atlas*. Some of the galaxies referred to in the classification are reproduced in this chapter. The classification symbols proposed by Hubble appear in parentheses next to the NGC numbers below each photograph.

ELLIPTICAL GALAXIES

Subclassification into ellipticity groups is made from the geometry of the projected image. The individual true ellipticities of the meridian sections are unknown except in the pure class E7 because the orientation of the principal axes to the line of sight is never known. The apparent ellipticities are expressed as 10 $(a\text{-}b)/a$, where a and b are the diameters of the major and minor axes, respectively. The observed classes range

*1961 Publication 618. Carnegie Institute of Washington, Washington, D.C. 20005. Reproduced by permission of the publishers.

from E0 to E7. No elliptical galaxies are known that are flatter than E7. Galaxies whose central sections *are* flatter invariably show an outer region of low surface brightness which resembles a thin, fundamental plane. Such galaxies are classed a $S0_1$ and are shown on page 4 of the illustrations.

The surface brightness of true E galaxies decreases smoothly from the nucleus, closely following the equation $I = Io\ (r/a + 1)^{-2}$. Here r is the nuclear distance, and a is a parameter which differs from galaxy to galaxy. The photographic images appear completely smooth with no breaks or inflections in the luminosity gradient.

The absolute visual magnitude of the brightest stars in E galaxies is about $M_v = -3.0$. The stars are red with international color indices of about 1.5. They are probably in the same stage of their evolutionary history as stars in globular clusters. E galaxies have no bright blue stars like those in the spiral arms of Sb, Sc, Irr, SBb, and SBc systems.

E galaxies represent an important stage in any theory of galaxies because conditions appear to be rather simple in them. The stars are old. They are in an evolutionary state which is fairly well understood. No new stars are being formed. There is no dust. There are no spiral arms. E galaxies do, however, have gas at very low densities. The evidence is the presence of emission lines in the spectra due to forbidden [O II] at $\lambda 3727$, and $H\alpha$ which is probably present whenever 3727 occurs. Most E galaxies show this emission.

It should be emphasized that the ellipticities of the images of E galaxies are apparent ellipticities only. They result from the projection of the true spheroid on the plane of the sky. A true E5, for example, can appear in projection as anything from E0 to E5, depending on the orientation of the symmetry axis to the line of sight.

S0 GALAXIES

S0 galaxies have symmetrical forms which are flatter than E7 but which show no spiral structure and no trace of bars. The characteristic features are a bright nucleus, a central lens surrounded by a faint and sometimes extensive envelope, and, in the later stages of the section, circular absorption lanes.

S0 galaxies appear to form a transition between the E galaxies and the true spirals. It is well established observationally that no elliptical galaxies exist whose fundamental planes have a larger flattening ratio than 1 to 3, whereas all spiral galaxies have fundamental planes whose flattening ratios *are* larger than 1 to 3. These results suggest that there exists a critical eccentricity of the central section of E systems. Some type of dynamical instability seems to begin in galaxies whose eccentricity is greater than this critical value. Galaxies that have angular momenta greater than the critical value tend to take up forms whose central sections are more eccentric than E7, and, for these, the instability causes matter to slough off the central regions and to spread out into a thin, disk-shaped structure which constitutes the fundamental plane. The resulting equilibrium form then consists of a dense nuclear region embedded in a thin disk or envelope. This may not, of course, be a correct description of the formation of the S0 forms, but the observed features of the class can be accounted for in this way. At any rate, the S0 appear to be important in any scheme of classification. They have the highly flattened fundamental plane of the true spirals together with the amorphous texture and the absence of dust and spiral arms of the E galaxies.

The transition from E to S0 is smooth and continuous. The gradual merging of the classes introduces uncertainties in the classification of borderline galaxies. The division between E and S0 is a matter of definition and is made on the basis of the presence or absence of an outer amorphous envelope or thin fundamental plane surrounding the nuclear regions.

The S0 are divided into three major subgroups, $S0_1$, $S0_2$, and $S0_3$. These subgroups represent extension along the sequence from early to late forms. The $S0_1$ is the most interesting from the dynamical standpoint because it represents the earliest of the class and is the nearest to the E forms. It is here that the first vestige of the thin fundamental plane appears. The distribution of light across the $S0_1$ is continuous, with no trace of absorption lanes or patches and no trace of structure. Except for the flatter intensity decrement, this type resembles E nebulae.

$S0_2$ galaxies show the same pattern of nucleus, lens, and envelope as the $S0_1$, but there is some structure in the envelope. On simple inspection, the lens appears to be surrounded by a relatively dark zone followed by a more luminous one. This description, however, represents a subjective interpretation of faint images and is deceptive. On photometric tracings, the luminosity gradients fall continually outward from the nuclei. The dark zone corresponds with a region in which the rate of fall first increases, then slows, and finally approximates the normal value. The form of the curve might be accounted for either by a circular ring of partial obscuration in the middle of the envelope or by a concentration of luminosity near the boundary. Either explanation indicates a segregation of material.

The weak absorption lane is silhouetted against the bright background of the nucleus in $S0_2$ galaxies seen edge-on. Because the ring is internal, the outer parts of the disk are unaffected and appear as bright ansae. Although the effect is rather inconspicuous because the internal obscuration is not opaque, it can be seen in NGC 4215, which is the best edge-on example. Here the nucleus and the outer regions of the fundamental plane are separated on each side by a zone of lower luminosity.

Galaxies in the $S0_3$ subgroup exhibit a structureless envelope similar to that of the $S0_1$, but a sharp, narrow, absorption lane is found within the lens. The lane is in an arc concentric with the nucleus. The structure is a later and further developed form of the $S0_2$. The arc appears to lie in or near the fundamental plane, and it cannot usually be traced over a complete circle. The photographs suggest that the arc is buried in the lens, and is overlain by luminous material. Since the most conspicuous portions of the arc generally parallel the major axes of the projected images, the incompleteness of the circles might be accounted for by the orientation effect.

The transition between normal S0 and Sa galaxies is gradual. The absorption rings in $S0_2$ and $S0_3$ galaxies begin to deviate from circular patterns and take on a tightly wound spiral form. Two transitional galaxies are illustrated, 2855 [$S0_3$/Sa(r)] and 5273 [$S0_2$/Sa(r)]. The circular pattern in 2855 is broken, and absorption lanes begin to wind outward through the envelope.

Sa GALAXIES

Normal spiral galaxies emerge from the S0 sequence with the appearance of noncircular absorption patterns and definite spiral arms. Although normal spirals show a wide diversity of form, they can be placed in a rough progressive order by three classification criteria: (1) the openness of the spiral arms, (2) the degree of resolution of the arms into stars, and (3) the relative size of unresolved nuclear region.

On this basis, the spiral sequence is divided into the three subsections Sa, Sb, and Sc. There is a smooth transition between the sections, and the boundaries are somewhat arbitrary except at the two ends. In general terms, the type S0, from which the spiral sequence emerges, is characterized by a structureless lens and an extended envelope ($S0_1$) or by circular patterns of obscuration ($S0_2$, $S0_3$). The first true spirals are called Sa. They show tightly wound spiral patterns of obscuration, and they may or may not have tight spiral arms of luminous matter. The arms are invariably smooth, with no resolution into stars.

The assignment of galaxies to the Sa, Sb, or Sc type is based here primarily on the characteristics of the arms. This does not mean that Sa galaxies do not exist with large amorphous central regions devoid of dust and spiral structure; all Sa galaxies of the 1302 type have this feature. We only wish to point out that a large amorphous central region is not a prerequisite for Sa galaxies.

SUMMARY

1. Spiral arms first emerge in the Sa galaxies.

2. The Sa type covers a fairly large section of the spiral sequence. The early Sa have ill-defined arms which are smooth in texture and are tightly wound about the amorphous nuclear regions. The principal classification features are the nearly circular pattern of the arms and the lack of resolution in the arms. Late Sa have prominent dust lanes (e.g., 3623, page 11) and a suggestion of lumpiness in the texture of the arms, showing the start toward resolution. Resolution begins in the early Sb systems and becomes "complete" in the Sc. But the arms are still tightly wound in the late Sa and form nearly circular patterns.

3. The nuclear regions of Sa galaxies can be either large (2775, page 10) or small (4866, page 11). There is not a unique connection between the relative size of the nuclear region and the arm pattern, but there does appear to be a strong correlation between the tightness of the spiral pattern and the lack of resolution of the arms into stars and H II regions.

Sb GALAXIES

The differentiation of an intermediate class Sb between the early and late stages of the spiral sequence is necessarily arbitrary. The two ends of the sequence, Sa and Sc, are easy to define. The Sa spirals have incipient spiral arms which are usually smooth in texture and are tightly wound in nearly circular patterns about an amorphous central region. The Sc spirals have highly branched, well differentiated arms that are not tightly wound around the nucleus and are well resolved into stars and H II regions. There is much dust in the arms. The nuclear region is usually small and inconspicuous. The large number of galaxies falling between the two extremes are called Sb. Class characteristics are formulated in terms of intermediate criteria between the Sa and Sc. Although the separation may seem vague in principle it works well in practice.

Sc GALAXIES

Dominance of multiple spiral arms and small amorphous nuclear regions are the characteristics of Sc galaxies. These features and the high degree of resolution of the open, branched-arm system make Sc galaxies easy to recognize. The detection of resolution in spiral arms depends on distance and on telescope size, but in practice the 60-inch, 100-inch, and 200-inch telescopes give photographs of sufficient scale so that either individual knots (H II regions and association of stars) or a general lumpiness in the texture of the arms can be detected in galaxies with redshifts as large as 15,000 km/sec. All the near-by bright galaxies are within this limit.

The Sc galaxies exhibit great diversity in form. At least six major subgroups can be distinguished. These groups in general represent development along the sequence of classification from early to late Sc. The largest subgroup of Sc galaxies resembles M101 (page 27). The arms are thin, multiply branched, and loosely wound about a very small nuclear region. They can be traced through about one revolution, although in some galaxies, such as NGC 5364 (page 32), segments of the main arms make nearly two complete turns. But this is rare. Although most Sc galaxies have multiple arms, close inspection of the photographs shows that only two predominant arms emerge from opposite sides of the nuclear region. They wind outward for nearly half a revolution, at which point they branch into many individual segments which continue to spiral outward to form the multiple structure.

SUMMARY

1. Sc galaxies have very small amorphous central regions. The spiral-arm pattern dominates the structural form on photographs taken in blue light.

2. Nearly all Sc galaxies have multiple spiral arms loosely wound around the central regions. Two principal arms can generally be traced almost to the very center of the system. Multiple branching creates new

arms from the two "parent" arms. The branching occurs after the parent arms have wound outward for about 90°.

3. The arms show a very high degree of resolution into knots, which are mostly H II regions and associations of stars.

4. There is much dust in the spiral arms; usually it lies on the inside of the luminous arcs. Dust arms are traceable into the nuclear regions (see the illustrations of M100, M51, M101, and NGC 628 on page 31).

5. There is great variation of structural form along the section from early Sc to late Sc. Six major subgroups are recognized on the basis of the cleanness of the arm separation. (There is very clear arm separation in the M101 group, and very poor in the NGC 2903, M33, and NGC 4395 groups.) The division into subgroups, although perhaps somewhat artificial, illustrates the progression along the sequence. The progression is shown in figure 2.

6. The transition from late Sc to Irr is smooth. NGC 4395-4401, NGC 45, and NGC 5204 are transition examples.

IRREGULAR GALAXIES

Irregular galaxies comprise 2.8 percent of the galaxies north of $\delta =$ -15° brighter than $m_{pg} = 13.0$. Irregular systems divide sharply into two groups. Galaxies of the first group are highly resolved into luminous O and B stars and H II regions. The systems show no circular symmetry about a rotational axis. Prominent spiral structure is missing. These galaxies form a continuation of the late Sc forms of the NGC 4395 type (page 37). The type examples are the Large and Small Magellanic Clouds. Galaxies of the first type are called Irr I. Galaxies of the second group also show no rotational symmetry. The photographic images are completely smooth in texture, show no sign of resolution into stars, and are often crossed by irregular absorption dust lanes and patches. These systems are called Irr II. The type example for the group is M82 (page 41).

BARRED SPIRALS

The class of galaxies characterized by a bar across the central regions was first recognized by H.D. Curtis, who assigned to it the provisional name "Φ type" spirals. Later the more convenient term "barred" spirals, represented by the symbol SB, was suggested by Hubble and has been universally adopted. The SB class includes 15 per cent of the brighter galaxies. The members do not differ sytematically from normal spirals either in luminosity, dimension, spectral characteristics, or distribution over the sky.

In describing the barred spirals it will be convenient to differentiate (1) the nucleus, (2) a region concentric with the nucleus which has the shape of a convex lens seen in projection, (3) a ring of luminous matter which when present lies on the rim of the lens and in the progression of the (r) subtype develops into the spiral arms, and (4) the envelope or outer region of the galaxy. In addition, there is always present the characteristic feature of the class, a more or less well defined luminous bar extending centrally across the lens and terminating at its rim or on the innermost coils of the spiral arms.

SB0 GALAXIES

The SB0 class represents a transitional stage between the elliptical galaxies and the true barred spirals. In this respect, the class is similar to S0 in the sequence of normal spirals. The boundary of the SB0 is defined on the E side by the presence of the characteristic bar and on the SBa side by the absence of spiral arms. Features common to the entire class are a nucleus, a lens of lower intensity concentric with the nucleus, the bar which usually terminates on the periphery of the lens, and occasionally an outer envelope and/or an external ring. The bar can be either very indistinct and difficult to recognize or well developed and prominent. The development of the bar is interpreted as a progression along the sequence from early to late SB0, and is recognized in the classification by the subdivision into three groups $SB0_1$, $SB0_2$, and $SB0_3$.

SUMMARY

1. There are three subtypes, $SB0_1$, $SB0_2$, and $SB0_3$, which differ from one another in the characteristics of the bar.
2. The bar of $SB0_1$ galaxies is a broad, indistinct region whose surface brightness is higher than that of the surrounding lens. It is inclined at a random angle to the major axis of the projected image.
3. The bar of $SB0_2$ galaxies does not extend completely across the face of the underlying lens. There are two diametrically opposite regions of enhanced luminosity on the rim of the lens which, together with the nucleus, constitute the bar.
4. The bar of $SB0_3$ galaxies extends completely across the face of the lens. It is narrow, well defined, and bright.

SBa GALAXIES

The SBa is the earliest of the true barred-spiral forms. There is a spread along the sequence from early to late SBa which is recognized by the developing characteristics of the spiral arms. The arms of the early SBa are poorly defined, smooth in texture, broad, and fuzzy. They are usually coiled in a nearly circular pattern around a central region which itself closely resembles the nucleus, lens, and bar of the $SB0_3$. The earliest of the SBa is NGC 936 [SBa(s)], which is now shown in the atlas because of the difficulty of reproducing the very faint arms. The earliest SBa illustrated is NGC 7743 [SBa(s)] (page 44). Here the arms are insignificant relative to the nuclear regions; they are smooth in texture and closely wound. There is no trace of resolution. In the later SBa, the arms are relatively narrow and well developed, and some show beginning traces of resolution (NGC 3185, page 43; 175, page 43). The arms here are tightly wound and form nearly circular arcs that appear in projection as elliptical loops.

SUMMARY

1. The characteristic bar is prominent and smooth in texture with no trace of resolution into knots or stars. No dust lanes are present in the bar a in SBb (s).

2. Spiral arms first appear in the SBa. They are usually closely coiled about the central lens and bar; they are often faint and inconspicuous, and usually smooth in texture although traces of partial resolution begin in the very late SBa.

3. The arms can either begin tangent to an internal ring [(r) subtype], or can spring from the ends of the bar [(s) subtype]. This division into (r) and (s) groups is not prominent in the SBa but is a dominant feature of SBb and SBc galaxies.

SBb GALAXIES

The classification criteria for the SBb are intermediate between those of the SBa and the SBc. They are based primarily on the openness and the degree of resolution of the spiral arms, and on the complete lack of resolution of the bar. The SBb galaxies break rather sharply into two subgroups, depending on whether the spiral arms start tangent to an internal ring [the (r) subtype] or spring from the ends of the bar [the (s) subtype]—the same two basic subgroups that have been discussed throughout this atlas.

SUMMARY

1. SBb galaxies have a well defined bar structure which is smooth in texture with no hint of resolution into stars.

2. There are two subgroups. The spiral arms of the SBb(s) group spring from the end of the bar at right angles. Two straight dust lanes in the bar turn sharply at the end of the bar and follow the inside of the spiral arms. NGC 1300 is the prototype. The spiral arms of the SBb(r) group start tangent to an internal ring on which the bar terminates. No dust lanes are present in the bar. NGC 1398 and NGC 2523 are prototypes of this group.

3. Transition objects exist between the (s) and (r) subgroups (e.g., NGC 4593 and NGC 4548), and between the SBb(s) and Sb type (e.g., M83, NGC 1097, and NGC 6951).

SBc GALAXIES

The characteristics of SBc galaxies are (1) the high degree of resolution of the bar and of the spiral arms into knots and luminous lumps, and (2) the openness of the spiral arms.

Six SBc galaxies are shown on page 49 of this atlas. As in the SBb, the SBc can be separated into the two subgroups (r) and (s), although the *pure* ringed type is not common. Most central rings are broken into several segments. An example is NGC 1073 (page 49), where the ring is not complete but breaks about 30° from either end of the bar. Of the six galaxies illustrated, three are of the pure SBc(s) type and three are of the combination type [SBc(sr)].

Note the high degree of resolution into stars of NGC 7741. Both the arms and the bar are resolved into knots. This feature of the resolution of the bar is unique to the SBc type. The bars in SBa and SBb galaxies are smooth in texture, with no hint of resolution.

The First Evolutionary Stage —
The Origin

Formation of Galaxies

A general pattern for the formation of galaxies from newly created matter was presented in Chapter IV. That scheme described the origin of a number of second generation galaxies from the original galaxy, but the description is applicable to the origin of any galaxy. The prime reason for the formation of new generations of galaxies has been the limitations of the effective ranges of the gravitational forces from the central regions. As more and more stars were formed farther and farther away, the gravitational forces from the center had less and less influence on them. Eventually, at great distances, the local forces took over; the stars became independent of the central region and moved away to form new galaxies.

In the same discussion we referred to "galactic supply regions" which were envisioned as roughly spherical regions within the range of the gravitational forces of the central region. It was then indicated that the periphery of the "galactic supply region" of the established galaxy was a neutral zone, from where the newly formed stars would move either inward toward the mother galaxy or away; a large number also stayed around and joined together to form patches of star-clouds. The fate of these clouds will be discussed presently. The stars that followed the expanding pattern of creation of matter and moved away radially, became attracted to each other in groups and then formed converging, galactic-type clusters. As usual, the convergence was caused by the gradual increases in the strengths of gravitational forces associated with the process of aging. Also, it should be mentioned that the motions of these stars were unlike those of the contemporary galactic clusters in that they were not subject to the shearing effects of forces from a central galactic region.

The Second Evolutionary Stage — The Adjustments

It took a considerable time for a converging galactic cluster to develop into an embryonic nucleus at distances far away from its point of origin. Also, during this period many stars had been formed in the new "galactic supply region," with the ones along the path of the cluster joining the group. The formation of the nuclear core occurred when the stars pulling closer together formed a system of many stars with complicated orbital patterns. With time the core of the aggregate which contained the oldest stars in the group became more compact. But new stars were being formed continually; they were pulled in toward the center and accumulated on the outside regions of the growing center. By this time the embryonic nucleus was somewhat similar to a globular cluster (Figure 29), though the peripheral regions did not end so abruptly; here the concentration of stars in the new structure gradually diminished with distance from the center. The central structure continued to grow in size, and as it became larger and older its gravitational influence became greater as well.

Factors Influencing the Structural Forms of Galaxies

Three principal factors have influenced the shaping of galactic forms. These are: (1) the amount of material that was available to the galaxy; (2) the number and masses of the initial stars in the nucleus and the rate of increase in their densities as they aged; and (3) the rate of rotation of the nucleus. These are interdependent factors; a change in one had profound effects on the others.

1. Availability of Material

Since the growth of a galaxy is totally dependent on the number of stars it can accumulate, the size of the "galactic supply region" was the prime factor in determining the final size

Figure 29. THE GLOBULAR CLUSTER M13 IN HERCULIS (*Hale Observatories Photograph*).

of the evolving galaxy and the degree of differentiation in its structural form. But the size of the "galactic supply region" was regulated by the total mass of stars in the central region, and by the number of galaxies in the neighborhood and their closeness to the growing galaxy. *Competition has always been the very essence of the universe*, and true to this rule, galaxies have competed for all the matter they could attain; the closer a galaxy was to a large evolving galaxy, the less was its chance to reach maturation. The results of this type of competition may be seen quite clearly in the Stephan's Quintet (Figure 30). Here, five galaxies have competed for the material which was probably sufficient for the formation of one large galaxy. Each has gathered as many stars as it possibly could, but because of a shortage of supply each has evolved into an incomplete galaxy. However, the accumulation of stars is still in progress as evidenced by the presence of star-streams between the galaxies. At present these star-clouds are in gravitationally neutral zones, but as the stars and the galaxies grow older, the star-streams will be attracted to the nearest galaxy and become absorbed. The possible mode of division of the star-clouds may be seen in the thin streamer curving over the double galaxy; it has split in two parts, and each section will probably join the closest galaxy.

Another example of the results of galactic competition is found in our neighbor, the Andromeda galaxy (Figure 31). NGC 224 (M31) has two companion galaxies: the spheroidal (NGC 221), the one close to the nucleus; and the elliptical one (NGC 205) at a considerable distance. Both of these dwarfed galaxies have had to compete with Andromeda for supplies and as a result their growths have become arrested. Not much can be determined about the evolution of NGC 221, but the elliptical NGC 205, classified under SO_1, has rudiments of primary arms. This indicates that while in competition with its huge neighbor it could not accumulate enough stars to grow to its full potential size.

The speed of aggregation of stars to form an incipient nuclear core has probably been the most important factor in preventing the formation of other nearby competitive nuclear structures. Naturally, the rate of pulling together was related to the number, the masses, the rate of energy production, and the

Figure 30. THE STEPHAN'S QUINTET (*Lick Observatory Photograph*). The degree of development of each galaxy has been dependent on the number of stars and star clouds it could attract from its surroundings.

Figure 31. THE ANDROMEDA GALAXY NGC 224 AND SATELLITES NGC 205 AND NGC 221 (*Hale Observatories Photograph*).

manner of movements of the stars in the converging cluster. If they came together to form the central core at a relatively fast rate, they would accumulate most of the newly formed stars within a large radius and thus reduce the chances for the formation of potential competitive centers within close ranges. For the same reasons, the clusters with small numbers of massive stars could not gather enough members to grow into mature galaxies.

Two principal types of incompleted galaxies, the young and the deprived, are of great value in tracing the evolutionary courses of galaxies. However, at present there is not enough information for the proper differentiation between these two groups. The possible clues for distinguishing one group from the other seems to be in the degree of compactness of stars, and the closeness of the suspected dwarfed galaxies to their neighbors.

2. Influence of Number and Masses of Stars in the Core

The transformation of a galactic cluster to a nuclear core was a very slow process, probably taking billions of years for its completion. In the beginning, the stars of the clusters were moving more or less parallel, but as they aged and their mutual gravitational attractions increased, they gradually converged toward a point far ahead in space. The degree of this convergence was a function of the masses and the number of stars in the cluster, the same factors mentioned in the previous section. This resulted in the formation of a central core. The convergence of the stars of the cluster and the following slow aggregation of newly formed stars around the core resulted in many double, triple and multiple stars. Then as the stars aged and further contractions occurred, the motions within the system grew more complex. This process continued until an equilibrium became established and the aggregate was rotating as a unit. The rate of rotation of the aggregate in turn determined the pattern of accumulation of the newly formed stars and thus the structural form of the galaxy.

3. Effects of Rate of Rotation of Nucleus

The growth of a galaxy does not take place in interrupted stages; it is always a product of continual accumulation of new stars, whose relative motions determine the final shape of the ġalaxy. In general, the motions of the newly formed stars are the outcome of orbital motions around the new galaxy's nucleus and the very slow inward motions toward the center as the orbital radii shorten with age. If the conditions of accumulation predicate the development of little or no rotation of the nucleus, each star will develop an independent orbit and the growth of the galaxy will be uniform in all directions, resulting in an E_0 - E_3 type of elliptical galaxy. But if the nucleus rotates at an effective speed, the orbital motions of the stars become influenced by the dragging effects (this process will be discussed later) of the central structure and follow the rotation. They will thus develop orbital motions around the equatorial plane and produce primary and secondary arms.

It should be remembered that in the classification of galaxies, no distinct lines have been drawn between E_7 and SO_1 classes; in fact it has been specified that the transition from E to S0 is smooth and continuous and the division is a matter of subjective definition. In order to emphasize this point, a number of edge-on photographs of galaxies of different classes have been placed next to each other in Figure 32. It may be seen that the extent of the formation of the arms is to some degree inter-related with the flattening of the nucleus at the poles (an indicator of the rate of rotation).

Evolution of Elliptical Galaxies

So far the discussions have been concentrated on the relationships between elliptical and spiral galaxies. Now we shall follow the evolutionary trends of the elliptical galaxies while keeping in mind that they are related to spiral galaxies. Whereas the recognition of evolutionary stages of spiral galaxies is not especially difficult, many problems exist in placing the elliptical galaxies in their proper evolutionary positions; with the latter it is even hard to separate the mature galaxies from immature ones. The probable criteria for judging age in elliptical galaxies consist of the degree of their compactness, the presence or absence of large numbers of globular clusters, their close proximity to much larger galaxies, and the presence of certain types of pulsating stars within the galaxy. In the studies of evolution-

NGC 4486 (EO pec)

NGC 4406 (E3)

NGC 4697 (E5)

NGC 4684 (SO₁)

NGC 5907 (Sc)

NGC 3115 (E7/SO₁)

NGC 4244 (Sc)

Figure 32. RELATIONSHIP BETWEEN ELIPTICAL AND SPIRAL GALAXIES (*Hale Observatories Photographs*). Most probably the differences in the structural forms are caused by the differ-

ary stages of elliptical galaxies, there is another problem: the shortage of sufficient numbers of photographs of truly elliptical ones. Statistically, this is as to be expected. If the rate of rotation of nuclei has been the prime factor in determining the shape of galaxies, then elliptical galaxies should form only a small percentage of the galactic population; the probabilities favor the development of some spin during the origin.

Figure 33 contains three photographs which probably represent different stages in the evolution of elliptical galaxies. By the evolution of elliptical galaxies we mean the different stages from infancy to maturity of galaxies evolving from nuclei with little or no rotation. Two of the galaxies, NGC 185 and NGC 205, belong to the Local group. The advantage of including these lies in the availability of their highly resolved photographs which give us the means of estimating their evolutionary maturity. NGC 185 seems to be the youngest photographed elliptical galaxy. This estimate is based on the degree of compactness of stars surrounding the nucleus and the presence of a dust cloud in the lower left corner of the central region. Most probably the cloud is on the outer regions of the galaxy with the nucleus providing the luminous background. The presence of such a cloud is an indication that the nearby stars have not as yet depleted most of their atmospheric material. NGC 205, as mentioned before, is probably an arrested galaxy which, if Andromeda had not taken a large portion of its supply, would have grown into a slowly rotating spiral galaxy with miniature arms. It also contains some patches of dust clouds. Its age ought to be approximately the same as M31 (Figures 31 and 50).

NGC 4486 is probably much older than the above two galaxies. There are a number of reasons for this estimate: it is compactly packed, even on the outside. The central region is very large; this means most of the stars have high densities and have become components of the nucleus. There are fairly large numbers of globular clusters surrounding the galaxy, an indication of complete maturity, and there has been an explosion in the center of the galaxy (Figure 53) which, as we shall see later, is an indication of the presence of aged, dense stars in the central region.

NGC 185 (E pec)

NGC 205 (E pec/SO$_1$)

NGC 4486 (EO pec)

Figure 33. THE EVOLUTIONARY STAGES OF ELLIPTICAL GALAXIES
(*Hale Observatories Photograph*).

In addition to the above, we are presenting the photograph of an incomplete E galaxy, Leo II System, in Figure 34. In the *Atlas*, Leo II System has been classified as a dwarf-elliptical. It does not have a recognizable nucleus, but contains a considerable number of RR Lyrae variables; the latter indicate its age to be about that of the primary arms of the Milky Way. The absence of a nucleus was probably caused by the sparsity of stars in the original galactic cluster. With no powerful gravitational forces in the center, many of the newly formed stars in the outer regions of the "galactic supply region" drifted away toward other sources of powerful gravitational attraction in the neighborhood. Its incomplete elliptical structure suggests the presence of relatively weak and probably spread-out sources of attractive forces in its central region.

Evolution of Spiral Galaxies

It is believed that spiral galaxies have all passed more or less through the same basic evolutionary process. The observable differences may be ascribed to a number of factors. These may be enumerated as follows:

1. The original rate of rotation of the nucleus.
2. The star supply available for the formation of nucleus, primary arms and the star-clouds that developed into secondary arms later on.
3. The local differences in the mode of distribution of stars. This had pronounced effects on the manner of formation of the secondary arms.
4. The position of the star-clouds (the beginning of spiral arms) relative to nucleus and primary arms.
5. The age of the galaxy.
6. The possibility of evolution becoming arrested by the shortage of star supply.

We shall first discuss the suggested evolutionary courses of spiral galaxies; then we shall divide galaxies in groups according to our approximate estimates of their rates of rotation, and observe the evolution of each group separately. In determining these courses we shall make use of the photographs from the *Hubble Atlas of Galaxies*.

Figure 34. A DWARF ELLIPTICAL GALAXY. Leo II System (Sculptor Type) (E) (*Hale Observatories Photograph*).

The proposed pattern of development of spiral galaxies has been represented through a series of diagrams in Figures 35-39. The initial stage (Figure 35) consists of the formation of the core of the nucleus from the largest galactic cluster in the region which had moved away from its mother "galactic supply region." The convergence of stars has been caused by the gradual increases in their gravitational attraction in response to the process of aging. Here, attention is drawn to the accumulation of stars by the growing nucleus along its path; this means stars were formed at relatively fast rates on the borders of the cone-shaped tunnel that the nucleus was producing along its course within the new "galactic supply region."

One of the conclusions derived from the previous discussions is that, in a spiral galaxy, the stars of the nucleus are the oldest and those of the spiral arms the youngest. In other words, star formation within a "galactic supply region" should have occurred around the incipient nucleus at an accelerated rate. But this concept contradicts another postulate in this book—that is, star formation has followed the pattern of creation of matter. The creation of matter was advancing radially in the form of an ever-enlarging sphere; hence the stars formed in a "galactic supply region" should be approximately the same age. The question is: how does one explain the discrepancy?

This paradox may be explained if we accept that an aggregate of massive stars moving within a space with the potentiality of star formation causes the speeding up of star formation in the vicinity of its course. The mechanism of this process may be attributed to the dragging effects on the frozen masses by the embryonic nucleus. The powerful gravitational forces from this center would drag the primordial frozen masses into antagonistic regions at faster rates than would have occurred without such an influence. This would have reduced the time needed for encounters and captures. In addition, the dragging would give the developing stars a linear motion in the same direction as that of the aggregate—that is, away from the center of the universe.

To return to the scheme for the formation of a spiral galaxy, the diagram in Figure 36 shows the beginning of the effects of the rotation of the nucleus on the manner of accumulation of stars. It is a rotating aggregate of massive stars with a relatively

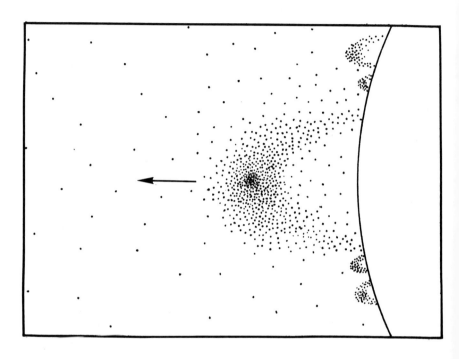

Figure 35. DIAGRAMMATIC REPRESENTATION OF THE INITIAL STAGES IN THE
DEVELOPMENT OF A NUCLEUS FROM CONVERGED STARS OF A
LARGE GALACTIC CLUSTER. The arc indicates part of the "galactic supply
region" of an older galaxy from where the galactic cluster originated. The four
small aggregates represent smaller galactic clusters that originated later from
the same region; they will evolve into globular clusters.

powerful gravitational attraction moving away from its point of origin; this movement speeds up the formation of new stars along its course and develops specific features, the most important of which are flattened poles and a bulged equator. In an evolving nucleus this distortion causes the equatorial region to have a more powerful gravitational attraction than the poles; consequently, as the dragged stars age, most of them move toward the equatorial regions. With each addition, the gravitational forces around the equator become stronger; this in turn causes the addition of more stars to that region. The result is the extension of the bulge outward in the shape of a disc. This gathering can be either in the form of two arms wrapped around the nucleus, or merely of an amorphous disc surrounding the central aggregate (Figure 40). Hubble has referred to these structural forms as "inner arms." We prefer to use the term "primary arms," which we believe is a suitable expression of their evolutionary order.

On the diagram in Figure 36 the primary arms have been formed and new stars are in the process of formation at relatively great distances above and below the poles of the young nucleus. The primary arms consist of stars drawn from the regions around the tunnel created by the rotating nucleus along its linear path. The stars close to the passing nucleus become a part of the central aggregate. The stars of the primary arms, being formed at greater distances, are younger than the nearer stars that have joined the nucleus.

At this stage, the rotating nucleus is surrounded by relatively young primary arms. As the developing galaxy moves away from its point of origin, it cuts the "galactic supply region" into two parts. By this time, the nearby newly formed stars have joined the evolving galaxy, and as a result a somewhat empty space have developed between the two regions above and below the poles (Figure 36). Because in the cut-off regions the gravitational attraction from the nucleus is ineffective in accelerating the rates of star formation, stars are formed at comparatively slow rates. In Chapter V it was shown that at certain stages in the evolution of stars most of the $+matter\text{-}antimatter$ reactions occur within mixtures composed of the two antagonistic gases. These reactions are similar but much more exaggerated

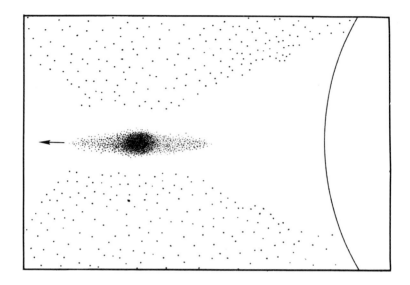

Figure 36. THE INITIAL STAGES IN THE FORMATION OF THE PRIMARY ARMS
AND THE APPEARANCE OF THE SEMBLANCE OF SECONDARY ARMS
IN A SPIRAL GALAXY. The stars of the nucleus and the primary arms have
been formed at an accelerated rate, catalyzed by the dragging effects of the
nuclear core. The stars of the spiral arms are in the process of being formed;
this involves encounters of antagonistic gases and the production of great
amounts of radiowaves. The pattern of generation and electromagnetic waves at
this stage of evolution is similar to the radio maps of quasars 3C 47 and 3C
249.1. (The Radio Maps of 3C 47 and 3C 2149.1 by G.G. Pooley and N.
Henbest. Obtained with the 5 km. radio telescope at the Mullard Radio As-
tronomy Observatory, University of Cambridge, England. *Monthly Notices of
the Royal Astronomical Society,* 169, 477 [1974].)

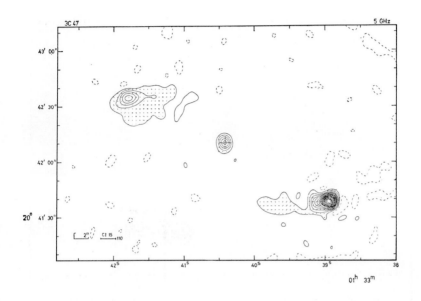

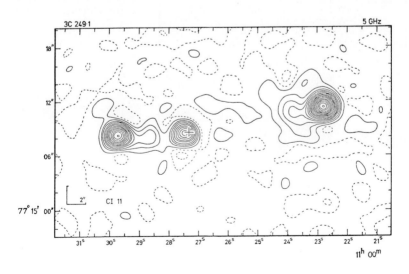

NGC 3329 (SO)

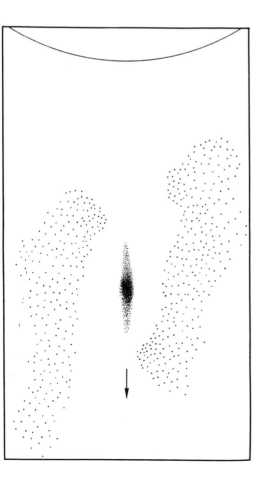

NGC 4215 (SO₂)

Figure 37. COMPLETION OF THE FORMATION OF THE PRIMARY ARMS AND THE BEGINNING OF THE ACCUMULATION OF THE STARS OF THE SECONDARY ARMS IN A SPIRAL GALAXY (*Hale Observatories Photographs*).

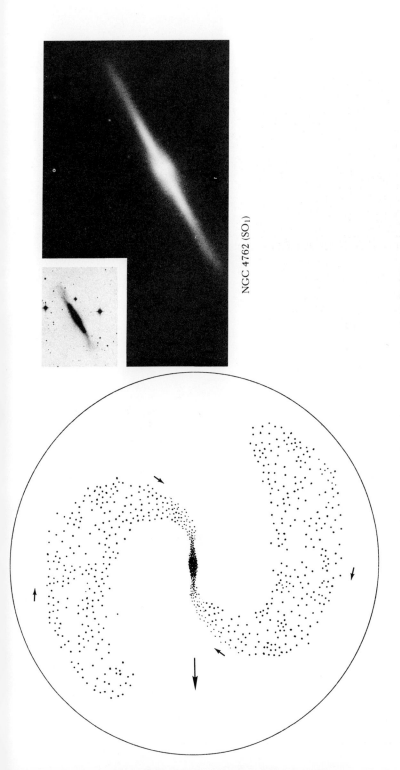

NGC 4762 (SO₁)

Figure 38. THE JOINING OF THE INCIPIENT SECONDARY ARMS WITH THE PRIMARY ARMS (*Hale Observatories Photograph*). The extension of the primary arms toward the incipient secondary arms is especially noticeable in the negative insert.

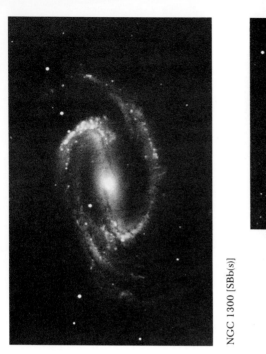

NGC 1300 [SBb(s)]

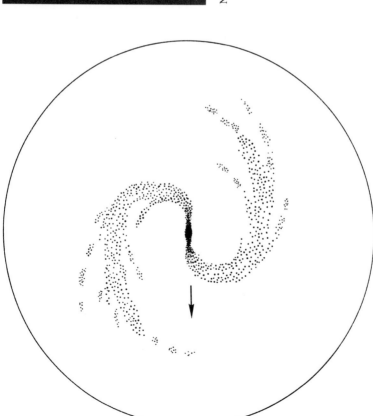

NGC 1073 [SBc (Sr)]

Figure 39. YOUNG SPIRAL GALAXIES (*Hale Observatories Photograph*). In most young spirial galaxies the secondary arms are twisted near the areas of attachment and then curve over and under the galactic plane.

NGC 5273 [SO$_2$/Sa(s)]

NGC 7743 [SBa (s)]

Figure 40. TWO PRINCIPAL FORMS OF PRIMARY ARMS (*Hale Observatories Photograph*). Both galaxies are very young; the stars in the arms are not as yet dense enough to produce compacted arms.

forms of reactions occurring in many nebulae in our Galaxy, from where considerable amounts of radiowaves are being generated. Consequently, at certain stages of their evolution the cut-off regions generate and emit great amounts of radiowaves. During this period the pattern of electromagnetic waves generation of the evolving galaxy consists of a young nucleus producing enormous amounts of high energy radiation with two oppositely located regions at a distance from the center emitting powerful radiowaves. This pattern is similar to the radio maps of the quasars 3C 47 and 3C 249.1 shown in Figure 36. If this description is correct, then these quasars may be regarded as young nuclei and the sources of radiowaves as the emerging spiral arms. This proposition becomes even more attractive if we also take into consideration the possibility that quasars are not as far away as their redshifts indicate (Chapter IV). It should, however, be remembered that this type of emission is a passing phase in the evolution of spiral galaxies. Once the formation of stars is about completed, instead of emitting powerful radiowaves, most of the emissions will consist of high energy radiation.

The Dragging Effect

We are now going to digress again from our general course of discussion in order to describe the probable manner of development of orbital motions of stars of the primary arms in an evolving galaxy. The theories in this book indicate that the orbital motion of a star along the galactic plane results from two inter-related motions, the already discussed motion caused by increases in gravitational attraction in the process of aging and the dragging effects of the motions of stars within and surrounding the rotating nucleus.

The basic movement of stars responsible for the development of orbital motions is the one associated with increases in the strengths of gravitational forces. Although this causes a shortening of distances between stars, its principal effect is the reduction of distances between stars and the nucleus. Whereas the increases in the attractive forces between two stars are on a one-to-one basis, those associated with the nucleus and a star result from one star being pulled in by increases in the strengths of forces emitted collectively by a multitude of stars. Thus as a galaxy evolves, although stars move imperceptibly but continually toward each other, the principal movements of the young stars are toward the nucleus. With newly formed stars this last motion eventually brings them within the effective dragging range of the orbiting stars that are moving in unison with the nucleus.

We have already discussed the way a nucleus may drag primordial frozen masses and newly formed stars along its linear path. The same effect also causes the orbital motions of stars in the arms. Here, we shall describe the simpler process of development of these motions among stars of the primary arms. Stars of the secondary arms go through the same steps later on, but those are via indirect routes. A rotating nucleus is made up of a large number of massive stars orbiting around the central core. Each star in the nucleus has its dragging influences on the stars farther away from the center. Thus when a star moves slowly toward the central region in response to the effects of increased gravitational attraction, it comes more and more under the dragging effects of stars that are being dragged by other stars closer to the nucleus.

The result is that the star develops an orbital motion which is in the same direction as the general direction of rotation of the nucleus, but not at a speed that makes it stay in the same position with respect to the dragging stars—it slowly falls behind. However, as it moves closer to the nucleus and the primary arms become more compact, the orbital motions of these younger stars become adjusted to those within the nucleus and move along their orbits in unison with the rotation of the central region. But this is not the case in the outer regions of the galaxy where stars are not so close to each other. This is well illustrated in our own Galaxy where, in the secondary arms regions, stars cannot keep up with the motions of their neighbors with shorter orbital radii.

Last Stages in the Formation of Spiral Galaxies

To go back to the discussion of the evolution of spiral galaxies, the next stage is diagramatically represented in Figure 37. Here the nucleus has become denser and more elliptical; also the stars in the primary arms have become more compact, and as a result the gravitational attraction at the outer edge of the galactic plane is very strong. While this has been happening in the central region, most of the matter in the "galactic supply region" has become converted to stars within the two regions above and below the evolving galaxy. It should be remembered that a "galactic supply region" is a roughly spherical space within the effective gravitational influence of the central region. This means that, though the local forces are more effective at great distances from the center and tend to hold the newly formed stars together in the form of star-clouds, the central gravitational attraction is dragging all of them along the linear course of the young galaxy away from the already formed older galaxies while remaining within the "galactic supply region." However, since the immature galaxy is rotating, it is establishing a balanced distribution of star clouds in the two regions above and below the nuclear poles; while being under the influence of the rotating primary arms, these clouds gradually join together to form star-streams. In addition to these adjustments, other changes take place in the mode of energy generation: most of the sources of weak +*matter-antimatter reactions* disappear, and at this stage instead of powerful radiowaves from the incipient secondary arms, most of the emission is in the form of heat, light and high energy radiation from the completely evolved stars.

The next stage of evolution is shown in the diagram of Figure 38. Here the process of aging has begun to show its effects: the primary arms have become more compact, stars in the developing spiral arms have come closer to each other, and the attraction between the star-streams and the outer regions of the primary arms has increased considerably. At this stage the stars in the star-streams are under the effects of gravitational attractions from two directions: first, the more effective and continually increasing lengthwise attraction along the incipient secondary arms in the direction of outer edges of the primary arms; and second, the less effective gravitational attraction coming directly from the nucleus. Usually the star-streams are far from the galactic center, are not directly over or below the nucleus, and do not show signs of being affected by the central region. In these galaxies the star-streams move slowly toward the primary arms at the nearest points and then join the galactic plane. As the evolution progresses, they follow the rotating motion of the galactic plane (Figure 39—NGC 1073 where the bar is the edge-on photograph of the nucleus and the primary arms). But occasionally the star streams are rather close to the central region; in those cases either the arms become bent toward the center (Figure 41) or are denuded of the more massive stars in the part of the arms nearest to the nucleus (Figure 39—NGC 1300). With regard to NGC 1300, attention is drawn to the absence of luminous stars in both arms in sections opposite each other and in line with the central region.

Members with Independent Orbital Motions

In addition to the already discussed parts of a spiral galaxy, two other groups of stars are present in the galaxy: the individual ones with orbits independent of the rotational motion of the main body of the galaxy, and the aggregates in the form of globular clusters. The orbits of the stars of the first group are elliptical and at angles to the galactic plane, a clear indication that they have originated near the nuclear poles. Most probably these stars were formed in regions not close enough to the poles for them to join the nucleus and yet not far enough to become members of the spiral arms. Their points of origin precluded them from coming under the dragging influences of stars of the galactic plane. The ages of these stars have been estimated from their properties to be approximately the same as those of the primary arms. This means that they came into being later than the stars of the nucleus but sooner than the ones that joined the secondary arms. As usual, their independent orbits developed and became stabilized in response to the process of aging and the gradual increases in the gravitational pull from the nucleus.

With regard to the globular clusters, three properties may be used in tracing their points of origin: (1) the pattern of the aggregation of stars in these units resembles those of the nucleus, indicating a similar type of origin; (2) globular clusters are also found among elliptical galaxies: this implies their formation to

NGC 3367 [SBc (sr)]

Figure 41. THE EFFECTS OF THE GRAVITATIONAL PULL OF THE NU-
CLEUS ON THE YOUNG SECONDARY ARMS FORMED CLOSE TO
THE CENTRAL REGION (*Hale Observatories Photograph*). Here the
arms are twisting over and under the nucleus. The galactic plane in this
photograph is at the edge-on position relative to our line of sight.

be independent of the rotation and the point of origin of the nucleus; and (3) the stars in these aggregates are approximately of the same age as the stars of the primary arms; this means their births have occurred after that of the nucleus.

These clues point to the possibility that, similar to the nucleus, these clusters originated within the old "galactic supply region" (Figure 35), and moved outward in the forms of small galactic clusters. But since they are younger than the nucleus, they ought to have been formed at a later period. By that time most of the stars had moved away and only a relatively small number were available for the cluster. In addition, having arrived later in the new "galactic supply region," they could not accumulate too many stars. However, their mode of aggregation to form globular clusters was the same as that of the nucleus. Then as the stars aged, the clusters developed independent orbital motions around the established and massive nucleus.

The Third Evolutionary Stage — Progress Toward Stability

The most effective stabilizing factor in the evolution of all galaxies is the continual increases in the strengths of gravitational forces caused by the process of aging. This phenomenon produces the imperceptible motions which have pulled the different components of the galaxies together from the beginning and will continue to do so until full maturity is attained. This very gradual transformation makes it difficult to separate the second evolutionary stage from the third, but then the same problem is present in all major evolutionary systems. In the case of galaxies, we have selected the completion of the formation and annexation of the secondary arms as the approximate end of the second stage. Since after this major event no more consequential additions to the bodies of galaxies occur, the evolution may be considered to have entered its third stage.

In a spiral galaxy the following three interdependent movements, all the outcome of the continual increases in gravitational attractions, contribute to its progress toward the best possible state of stability: the central region, namely the nucleus and the primary arms, becomes more compact as the galaxy grows older; in almost all cases the spiral arms fold; and finally, in many cases the secondary arms flatten—that is, through a slow untwisting type of movement they move from their positions in the polar regions toward the equatorial plane. The first type has been discussed repeatedly and no further explanation is needed here. The second and the third require more clarification.

Folding of Secondary Arms

The process of the folding of the spiral arms consists principally of the reduction of the orbital radii of the stars of the arms; these reductions also change the relative positions of stars within the arms. It was mentioned in Chapter IV that the rate of increase in the gravitational constant of a star is a function of its age and the rate of its energy production. Since the stars of the spiral arms in a galaxy are approximately of the same age, then the rate of increase in their gravitational attraction becomes a function of the rates of their energy production. Therefore, as the folding of the arms progresses, the orbital radii of the bright stars should become shorter at a faster rate than those with lower luminosities. These effects of non-uniform increases are recognizable in certain middle-aged galaxies where a considerable amount of folding of the secondary arms has taken place. The results may be seen in NGC 5194 and NGC 5248 (Figure 42); here many of the more massive stars of the secondary arms with their greater rates of energy production have increased their gravitational attractions at relatively fast rates and have been pulled in by the nucleus to the inner edges of the arms. The reverse of this phenomenon—that is, the accumulation of dust and small low-luminosity stars at the outer edges of the arms—is shown in Figure 43.

Flattening of Secondary Arms

The third movement, one of flattening, is a combination of two concurrently occurring motions. The first is the trailing of the motion of the galactic plane by the secondary arms, brought about by the dragging pull of the rotating primary arms. This

NGC 5248 (Sc)

NGC 5194 (Sc)

Figure 42. GATHERING OF MASSIVE HIGH-LUMINOSITY STARS AT THE INNER EDGES OF SPIRAL ARMS (*Hale Observatories Photographs*).

Figure 43. EXCESSIVE DUST CLOUDS AND LOW LUMINOSITY STARS AT THE OUTER EDGE OF A SPIRAL GALAXY (*Hale Observatories Photograph*).

effect is exerted at the points of the arms' attachments and all the way along their lengths. This dragging effect is most pronounced at the junction with the primary arms and becomes less and less effective with distance. The second is the very gradual compacting of the secondary arms caused by aging. This brings about nominal contractions of the widths of the arms and corresponding shortenings of their lengths. The first motion is the principal cause of flattening: the powerful gravitational forces of the galactic plane drag the arms and slowly bring them to the level of the primary arms. The second effect strengthens the bonds between the primary and secondary arms and thus assists in the flattening process. A fascinating photograph of the last stages of flattening of the secondary arms is presented in Figure 44. Here the secondary arms of the galaxy NGC 3190 have not as yet reached the level of the galactic plane in their untwisting motions. In the coming section, a number of evolutionary series will be presented, where it will be shown that the extent of folding and flattening provides useful criteria for the estimation of ages of galaxies.

It should be specified here that spiral arms do not always flatten in response to the aging of a galaxy. As already mentioned, the arms are under the effects of gravitational forces from two directions: the forces coming from the primary arms along the lengths of the secondary arms and forces from the nucleus pulling the arms toward the central region. Usually, the secondary arms are not formed directly over the center; rather, they twist at acute angles relative to the galactic plane (Figure 45); these secondary arms are not particularly affected by the nucleus and flatten out in time. But, occasionally, the secondary arms are formed directly over the nucleus at about right angles to the galactic plane; in these cases the dragging pull of the primary arms is ineffective and the atypical arms slowly fold over the nucleus (Figure 46).

Barred Galaxies

Barred galaxies are special cases of relatively young galaxies whose primary arms are more or less at the edge-on position with respect to our line of sight and their spiral arms at various

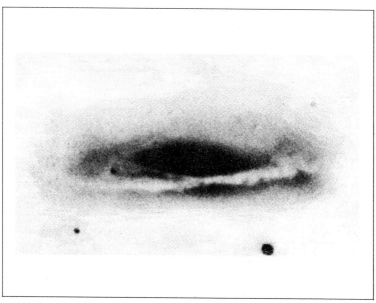

NGC 3190 (Sc)

Figure 44. THE LAST STAGES IN THE FLATTENING OF THE SPIRAL ARMS.
A negative photograph by Edwin Hubble, *Astrophysical Journal*, 97,
114 (1952).

NGC 210 (Sb)

Figure 45. TYPICAL POSITION OF THE YOUNG SPIRAL ARMS RELATIVE TO THE CENTRAL REGION (*Hale Observatories Photograph*). The arms are at considerable distances from the center and do not pass directly over the nucleus.

NGC 2685 (SO pec)

**Figure 46. ATYPICAL POSITION OF THE SPIRAL ARMS RELATIVE TO THE
NUCLEUS AND THE PRIMARY ARMS (*Hale Observatories Photo-
graph*).** The central region is at edge-on position relative to our line of
sight; the secondary arms are almost at right angles with respect to the
galactic plane.

angles relative to the galactic plane. That the bars are the edge-on photographs of the primary arms may be seen in the series of photographs shown in Figure 47, where four galaxies with different tilts demonstrate how the primary arms can give the impression of being bars. That they are fairly young is displayed by the absence of pronounced flattening and folding of the secondary arms. Some of the barred galaxies with secondary arms twisting over and below the nucleus will never become completely flattened out, but most of them will evolve into typical spiral galaxies.

Irregular Galaxies

As far as the irregular galaxies are concerned, they seem to be patches of star-clouds that came into being in the manner of the ones in the spiral arms, but without being near enough to a center to join a developing galaxy. Most probably they were formed in between two "galactic supply regions" where the attraction from the two centers neutralized each other and became ineffective. Then, as the two galaxies moved away from each other along their courses and away from the universal center (Chapter IV—Figure 5), these star-clouds drifted into space and became independent entities. For example, the isolated star-cloud between the two spiral galaxies in Figure 48 has the potential for becoming the center of attraction for smaller star-clouds and individual stars in its surrounding. After the two galaxies have moved away from each other, the cloud may evolve into an irregular galaxy.

Effects of Rates of Rotation of Nuclei on Evolution of Galaxies

In order to demonstrate the influence of the rate of rotation of a nucleus in the shaping of a galaxy, in this section we shall present the evolutionary stages of five groups of galaxies assembled according to the estimates of the rates of rotation of their nuclei. Unfortunately, the rates of rotation of the nuclei of most galaxies are not known. Here, tentative estimates have been made based on the sizes of nuclei relative to the size and forms

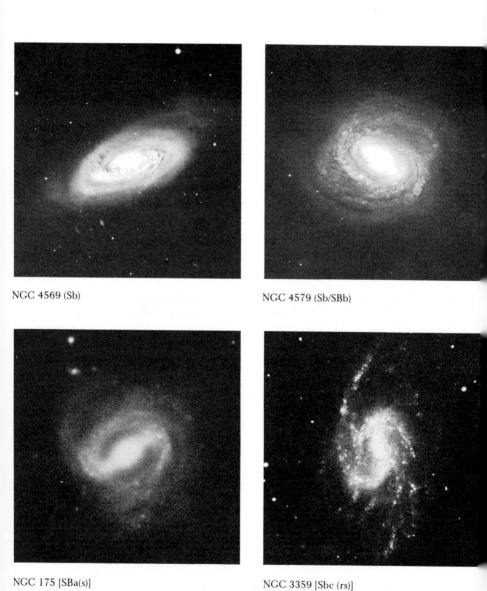

NGC 4569 (Sb)

NGC 4579 (Sb/SBb)

NGC 175 [SBa(s)]

NGC 3359 [Sbc (rs)]

Figure 47. FOUR GALAXIES AT DIFFERENT TILTS TO OUR LINE OF SIGHT (*Hale Observatories Photographs*).

Figure 48. STAR-CLOUDS IN GRAVITATIONALLY NEUTRAL ZONE BETWEEN TWO GALAXIES NGC 5432 AND NGC 5435 IN VIRGO (*Lick Observatory Photograph*). If conditions are right, the star cloud in the middle zone will remain independent of the two galaxies and by accumulating stars from its surroundir evolve into an irregular galaxy.

of primary and secondary arms. No claim is made for accuracy in these estimates. But until more precise determinations are made, this system of grouping may be used as a guide for following the stages in the evolution of galaxies.

A number of galaxies with specific characteristics have been arbitrarily divided into groups according to the estimated rates of rotation of their nuclei—that is, "little or no rotation," "slow," "medium," "fast," and "very fast." Evidently, a considerable number have rates of rotation in between the suggested rates; we should expect these to have characteristics intermediate between the two groups with closest rates. In addition, attention is drawn to the problems encountered in finding galaxies in their early evolutionary stages. These difficulties are compounded by the absence of specific characteristics in the very young galaxies; with these, reliance has to be placed on tenuous clues.

1. Little or No Rotation

These are the elliptical galaxies discussed in previous sections. The evolutionary stages of this group have been presented in Figure 33. As already explained, their lack of effective rotation has prevented them from developing either primary or secondary arms.

2. Galaxies with Slowly Rotating Nuclei

The main characteristics of the members of this group are large but slightly oblate nuclei with comparatively small primary and secondary arms. In these galaxies the slow rate of rotation has resulted in the following:

1. There has been little flattening at the poles.
2. The nucleus has grown large in the manner of elliptical galaxies. Also, a large number of stars have developed independent orbital motions and have not joined the galactic plane.
3. The large nucleus has pulled in many stars from the spiral arms regions; consequently, the secondary arms are relatively short and narrow.
4. The great gravitational pull of the nucleus has shortened the period of folding and flattening.

There are not many photographs of galaxies with "slowly rotating" nuclei. Of this group, NGC 4258 (Figure 49) appears to be the youngest. It has a comparatively large nucleus, two partially wound primary arms and a pair of diminutive secondary arms composed of thin star-streams. The brightest of the two spiral arms is attached to one of the primary arms, but the region of attachment of the second one is not observable, probably hidden under dust clouds. Though this galaxy has been placed in the "slowly rotating" series, it gives the appearance of its nucleus rotating faster than the other two members of this group.

The galaxies NGC 4800 and NGC 4594, the second and the third members of this group, are in the more advanced evolutionary stages. NGC 4800 appears to be the younger of the two, as shown by its incompletely folded secondary arms, but we do not have any means of estimating the extent of the folding of the arms of NGC 4594. However, the interesting point about these two galaxies is that one is a face-on and the other an edge-on photograph showing the two possible views of mature galaxies with "slowly rotating" nuclei. The edge-on picture also demonstrates the arrangement of stars with independent orbital motions and globular clusters around the slightly oblate nucleus.

3. *Galaxies with Medium-Rate Rotating Nuclei*

There are many good examples in this group. These galaxies have medium-sized nuclei, well-formed primary arms and, when in a mature stage, prominent secondary arms. Out of the available photographs, seven have been chosen and arranged in the probable order of advancement of their evolutionary stages (Figure 50).

In the galaxy NGC 2859, the fuzzy and indistinct accumulation of stars around the nucleus gives the impression of the primary arms being in their early evolutionary stage. No secondary arms are noticeable in this galaxy. In the next photograph, NGC 2681 shows the last stages in the folding of the primary arms. The well-defined arms have become tightly wound around the nucleus. The third galaxy in line, NGC 718, illustrates the early stages in the attachment of the secondary

NGC 4258 (Sb)

NGC 4800 (Sb)   NGC 4594 (Sa/Sb)

Figure 49. EVOLUTIONARY STAGES OF GALAXIES WITH "SLOWLY ROTATING" NU-CLEI (*Hale Observatories Photographs*).

arms to the primary arms. Here, two fuzzy spiral arms protrude from the wound-up primary arms. The secondary arms in this galaxy have not flattened out; the brighter one nearest to us is twisting over the central region while the fainter one is located under the young galaxy in the direction away from us. The reverse of the same arrangement in the secondary arms is seen in the next evolutionary stage, in NGC 1097. In both of these galaxies the secondary arms are twisted above and below the galactic plane and are at considerable distances from their respective nuclei. They may be considered as fairly young galaxies, but the degree of compactness of stars in the secondary arms indictates a younger age for NGC 718. In the next evolutionary stage, as represented by NGC 3504, the secondary arms have become considerably flattened and folded and are untwisting in the direction of the galactic plane.

As a galaxy in this group ages, the twists in the arms become transformed into smooth curves in the process of flattening. The graceful galaxy in Ursa Major, NGC 3031 (M81), clearly shows this type of transition. Here, the secondary arm in the upper left corner of the photograph is above the galactic plane with its opposite number extending below the body of the galaxy. The raised side may be compared to a football stadium where the plane of the secondary arm corresponds to the stands used by the spectators and the primary arms and the nucleus to the playing field. When this galaxy grows older and the flattening and the folding of the arms are completed, it will lose its youthful appearance and will look somewhat like the older and mature NGC 224 (M31), the last photograph of the series.

4. Galaxies with Fast Rotating Nuclei

The size of the nucleus relative to that of the arms may be used to differentiate the members of this group from the ones belonging to the galaxies with "medium-rate rotating" and the "very fast rotating" nuclei. The nuclei of galaxies in this group are smaller than the ones belonging to the galaxies with "medium-rate rotating" nuclei, but larger than those of the "very fast" ones. Another means of differentiation is the comparison of the primary arms. In this group the primary arms are somewhat

NGC 2859 (SBO$_2$)

NGC 2681 (Sa)

NGC 718 (Sa)

NGC 1097 [SBb(s)]

Figure 50. EVOLUTIONARY STAGES OF GALAXIES WITH "MEDIUM-RATE" ROTATING NUCLEI (*Hale Observatories Photographs*).

NGC 3504 [SBb(s)Sb]

GC 3031 (Sb)

NGC 224 (Sb)

smaller than the ones in the "medium-rate" group, but the difference is not pronounced enough to be of use for identification purposes. However, the "fast rotating" group can be easily identified from the "very fast rotating" ones; the latter have either miniscule or no primary arms.

Figure 51 contains five photographs of galaxies belonging to this group. The first one, NGC 23, seems to be the youngest. It has two primary arms in the process of folding and one visible secondary arm twisting under the galactic plane, probably in our direction. The opposite secondary arm is not observable, but there are traces of star-clouds in that region, indicating the second spiral arm in the process of formation, probably away from us. The second photograph in the series is that of a somewhat older galaxy, NGC 1073. Its secondary arms are well-developed and are twisted at sharp angles with respect to the galactic plane, the latter being at the edge-on position relative to our line of sight. The next evolutionary stage of this group is represented by NGC 6951. In this galaxy the primary arms are tightly packed and the secondary arms have nearly lost their angular twists. However, the distances of the spiral arms from the central region point to the still youthful state of this galaxy.

The last two galaxies, NGC 4321 and NGC 2903, show the mature stages of the members of this group. The relative ages of these two galaxies may be determined by the angles of pitch of their arms. An interesting comparison may be made between NGC 4321 and NGC 3031 (M81) of the "medium-rate" group. Both seem to be approximately in the same evolutionary stage; in both cases the secondary arms have not become completely folded or flattened.

5. Galaxies with Very Fast Rotating Nuclei

The galaxies in this group (Figure 52) may be differentiated from those of other groups in having small nuclei with either incompletely developed or no primary arms. The very fast rates of rotation prevent the nuclei from growing into large bodies, and small nuclei do not possess the powerful gravitational forces at their equators to draw in the proper number of stars for the formation of well-developed primary arms. In this group most

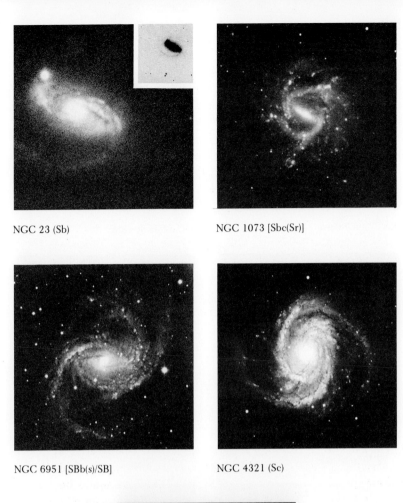

NGC 23 (Sb)

NGC 1073 [Sbc(Sr)]

NGC 6951 [SBb(s)/SB]

NGC 4321 (Sc)

NGC 2903 (Sc)

Figure 51. EVOLUTIONARY STAGES OF GALAXIES WITH "FAST ROTATING" NUCLEI (*Hale Observatories Photographs*).

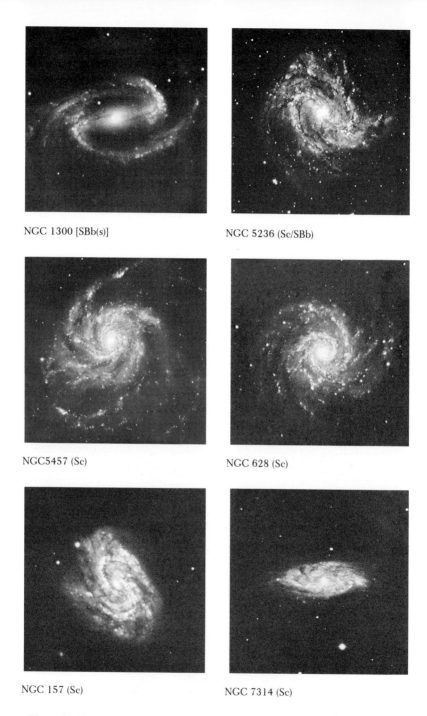

NGC 1300 [SBb(s)]

NGC 5236 (Sc/SBb)

NGC5457 (Sc)

NGC 628 (Sc)

NGC 157 (Sc)

NGC 7314 (Sc)

Figure 52. EVOLUTIONARY STAGES OF GALAXIES WITH "VERY FAST RO-
TATING" NUCLEI (*Hale Observatories Photographs*).

of the stars in the vicinity of the nucleus, instead of joining the embryonic primary arms, become the bases for the secondary arms. The cause for this arrangement is the stretching of the arms brought about by the late response of stars to dragging effects from fast-orbiting stars. Once a star comes under the dragging influence of other stars, it needs time to acquire its maximum velocity. Consequently, the faster the rates of motions of the dragging stars are, the longer will be their distances from the dragged star before the latter reaches its final speed. This means that the higher the angular velocity of the nucleus, the greater are the distances between the stars and the more stretched the arms.

An important characteristic of the older members of this series is the accumulation of most of the luminous stars along the sides of the arms facing the nucleus. This phenomenon has already been explained: the movements of the massive stars in the arms toward the nucleus are attributed to faster rates of increases in the strengths of gravitational forces caused by higher rates of their energy production. In these galaxies the greater distances between the stars of the arms allow the more massive stars to move inward without having to drag many of the smaller ones with them.

The first galaxy in this series, NGC 1300, has partially developed primary arms that are in the process of becoming attached to the secondary arms. The very fast rate of rotation and the stretching of the primary arms will probably prevent them from becoming wrapped up around the nucleus. Most probably these arms will become the bases of the secondary arms. The important characteristics of the secondary arms of this galaxy have already been explained.

The rest of the galaxies selected from this group have been arranged in the order of their evolutionary stages as judged by the degree of folding and the accumulation of bright stars at the inward edges of the arms. The stretched conditions of the arms have removed the importance of the flattening process in estimating the evolutionary stages of the older members of this group. The last member of this series, NGC 7314, gives the appearance of being an old galaxy. Here the arms have folded to such an extent that they have about lost their identities.

The Fourth Evolutionary Stage — Dissolution

Up to this point these discussions have covered the three principal stages in the evolution of galaxies. There is no reason to believe that the fourth and the last stage will not follow, and the slow deterioating process common to all evolutionary systems will not take place. True, we have records of explosions in the central regions of galaxies, such as in NGC 4486 and NGC 3034 in Figure 53. But these are occasional occurrences and galaxies do not necessarily have to be in their terminal stages in order to explode in the center. Since the normal pattern of all evolutionary systems is to come into being gradually and die slowly, we should expect the same to happen in most galaxies. In this regard, the proposed concepts in Chapter VI point directly to the way the dissolution will usually take place. It has been stipulated that the survival of the more massive stars depends on the utilization of material from their smaller companions. Therefore, the aging of a galaxy is accompanied by a continual decrease in stellar population density, an attrition process which does not usually end up in an explosion.

The understanding of the mechanism of an explosion in the central region of a galaxy would be of help in deciding whether or not explosions should be considered as the normal pattern of disintegration of galaxies. Explosions of extremely great magnitudes, such as the ones observed in some of the galaxies, can take place only if exceedingly large quantities of antagonistic material become mixed up at a fast rate. Most probably there are about equal numbers of +*matter* and *antimatter* massive stars in the nucleus. Under normal conditions they keep apart from each other through the effects of radiation pressures between them. However, if a binary among them turns into a super-supernova (a rare but possible occurrence among these high density stars), it can trigger a chain reaction in which the contents of antagonistic stars of the central region become mixed up. This can happen in any galaxy with dense massive stars in the center and does not necessarily indicate the normal manner of the demise of galaxies. Therefore, explosions should not be considered to be the usual pattern of the fourth stage in the evolution of galaxies.

NGC 4486 (EO pec)

NGC 3034 (Irr II)

Figure 53. EXPLOSIONS IN GALAXIES (*Hale Observatories Photographs*).

It should be pointed out that the most significant point regarding the fourth stage of the evolution of galaxies is not the mechanism of their dissolution, but the fundamentals associated with their births and deaths. A galaxy comes into being, lives its life and dies. During this period it converts most of its mass to radiation energies and emits about equivalent amounts of gravitational forces. In other words, it comes into existence from a state of "nothingness" and becomes dissipated into space in the form of energy, force and some residual matter. This is, in a true sense, the summary of what happens in the evolution of matter in the universe, the subject of discussion in the next chapter.

X

THE EVOLUTION OF MATTER

The discussions on creation of matter in the early chapters dealt with the formation of subatomic particles and the development of the state of orderliness of *force-, energy- and space-* primatons from the disorderly state of "nothingness." This transformation consisted of the continual production of pairs of symmetrical particles in association with the establishment of a constant linear velocity for primatons. On the basis of this and other related theories, it was concluded that the creation of matter has been and is a self-perpetuating phenomenon, advancing radially in the form of a somewhat spherical shell, forming stars and galaxies in its wake, giving the galaxies their away motions from the point of origin in the universe, and producing the phenomenon of the expanding universe. According to these theories, the creation of matter has continued ever since its origin, and the process is expected to continue as long as there are disorganized primatons in the universe.

It was reasoned that most probably neither the distances nor the rates of recession of the distant galaxies from the Earth are as great as the values obtained from the determinations of their redshifts. Consequently, the universe is believed to be smaller than the present estimates. We do not know the position of the point of origin in space, but the possibility exists for locating it by determining the true courses of motions of galaxies and by attempting to observe a number of old galaxies in the suspected region. The second part of these endeavors requires luck, as the areas through which such observations can be made are rather limited. Nonetheless, should we be fortunate enough to locate the old galaxies, we can then use their ages for estimating the approximate age of the universe. Most probably it will be found that the universe is much older than the accepted values of about twenty billion years.

Most of the concentrated matter in the universe is found in galaxies, but galaxies cannot be considered as the real participants in the evolution of matter. That role belongs to stars, within which the evoltuionary processes of matter are taking place. In this respect the principal function of a star is to emit both *force-* and *energy-* primatons. If we attribute the expansion of the universe entirely to the away motion caused by the pattern of expansion of the shell where matter is being created, then neither the produced repulsive energies nor the attractive forces have had or will have any significant effects on the movements of galaxies. This means that the forces and energies produced in the universe are neutralizing each other's effects and hence the sum total of all the energies radiated has been about equal to all the emanated forces.

This conclusion is supported by the facts that neither is the universe contracting nor are the galaxies moving ahead of the shell where the creation of matter is taking place. The isotropic cosmic background radiation, an indicator of the creation of matter beyond the very young galaxies and quasars, confirms the opinion. In this regard, attention is drawn to the statement in Chapter III where it was indicated that the Sun emits greater amounts of gravitational forces than radiation. This imbalanced relationship, however, does not apply to forces and energies emitted by galaxies. Although individual stars emit greater amounts of forces than energies, within a galaxy a large proportion of these forces is neutralized among the stars; hence they do not find their way into intergalactic space.

We can now summarize the basic steps in the evolution of matter. The process began somewhere in space, and, ever since, the disorderly primatons with "independent motions" have been converting to organized forms of subatomic particles which are then held together by force-primatons produced from space-primatons by the spin of these particles. On the other side of the picture is the inability of the produced symmetrical particles to remain apart forever; their eventual encounters result in their conversion to *energy-* and *force-* primatons which are then radiated into space (Chapter V). It has been shown that most probably the energy-primatons traveling through space encounter space-primatons and transfer part of their energies to them; the redshift caused by distance attests to the occurrence of this transfer. There is no reason to believe that the reverse of this phenomenon also is not taking place with the emitted force-primatons—that is, as they move through space, part of their force strengths are transferred to space-primatons. We may then conclude that, *most probably the energy-primatons and force-primatons emitted by stars, are continually exchanging their excesses through the medium of space-primatons, and both are shifting toward the state of complete balance between the rotational and linear motions of space-primatons.*

According to this interpretation, the evolution of matter consists of the conversion of disorderly primatons to matter, which in turn is transferred gradually to *force- and energy-* primatons destined to end up as space-primatons. *Consequently, matter may be considered as an intermediary between the two states of "non-existence" and hence transitory in nature.* However, since the length of this temporary state is exceedingly out of proportion relative to the lengths of our lives and all our other standards, it gives us the appearance of being endless and thus permanent. In order to dispel this illusion, all we have to do is to measure the loss of the mass of the Sun in one year, multiply it by the known number of stars and the estimated years of their existence and then, from the results obtained, attempt to answer the following two questions:

1. What has happened to all the masses that have disappeared since the origin of the universe?

2. What is going to happen to most of the remaining masses, supposedly composed of indestructible matter?

Such contemplations would make us aware of the passing nature of matter and the one-way manner of its dissipation into limitless space.

XI

THE EVOLUTION OF LIFE

We are now going to direct our attention away from the problems of the heavens to those of life on Earth, an outgrowth of the evolution of the solar system. In this chapter we shall concentrate on life's first and second evolutionary stages; the third one, the Darwinian evolution, is well studied and does not need any elaboration here. The final part of the chapter will consist of a brief review of the author's views about the position of Man in the evolutionary systems and his probable contributions to the evolution of matter.

The Preliminaries

Life may be defined as a system composed of multitudes of living units that have established a gradually changing state of equilibrium among themselves, the Earth and the Sun. A true living unit is then defined as an individually independent and integrated mutable organic system having a limited capability for adaptation to the environment and possessing the mechanisms for the intake and utilization of energy, either directly in the form of electromagnetic energies from the Sun, or indirectly from the products of photosynthesis (rarely by catalysis of inorganic compounds); this energy is then used for maintenance and growth, processes which usually lead to and result in self-replication.

The term "origin of life" has a tendency to give the impression that the beginning of life was a fairly rapid and consequential event during which the right molecules came together at the right time, in the right place, and under the right environmental conditions to produce the first self-replicating organism within a relatively short period of time. This misconception should be avoided for the simple reason that complex evolutionary systems do not originate in this fashion. An evolutionary system comes into being more or less on the trial-and-error basis—that is, probabilities favor its origin. If probability was the deciding factor, then the origin of life should have been a very slow, plodding process and, similar to the beginning of all major evolutionary systems, marked by numerous failures. *Such a pattern of development forces us to accept a long pre-cellular history during which the primitive self-replicating molecular aggregates laid the foundation for the emergence of the cellular organisms.*

The understanding of the pre-cellular evolution requires answers to a number of relevant questions of which the following are the most important: (1) how did the essential organic compounds become synthesized? (2) how did these molecules find each other and, in the absence of enzymes, join together to produce the self-replicating macromolecules? and (3) how did the metabolic cyclical systems form and become integrated with these macromolecules, thus paving the way for the units to evolve into living cells? These questions touch upon some very fundamental points regarding life, and we shall try to answer them as we progress with our proposals and hypotheses. But for the present we must prepare the ground by reviewing some of the pertinent facts in four separate fields: the role of solar energy in sustaining life; the relationship between self-replication and the evolution of life; the environmental requirements for the origin and continuation of life; and the inter-relationship of the evolution of the Earth with the origin and evolution of life.

Role of Solar Energy

The processes that keep the equilibrium of a living unit intact come from two recognizable and integrated systems, metabolism and reproduction, the former being the provider of material and energy for the consummation of the latter. Without resorting to the description of elaborate and involved biochemical schemes, the metabolic aspect of life may be described simply as processes that in almost all cases bring about the utilization of the energies from the Sun through numerous interwoven and complex mechanisms. Within any living unit these comprise cycles within cycles within cycles and so on, enmeshed into each other with almost flawless perfection. When one cycle completes its turn, it contributes its products toward the continuation of other cycles. The smaller ones are interlaced with each other to produce larger cycles; these in turn act as the components of still larger ones. Finally, the usual outcome of these actions and reactions is the completion of the largest cycle, reproduction. We shall discuss reproduction in the next section; in this part we are primarily interested in the movement of the solar energy through the systems that compose life on Earth.

The mechanisms that are instrumental in the storage and utilization of the Sun's energies operate in the form of a huge cycle composed of two inter-related processes. The first is photosynthesis; this brings about the storage of energy through the bonding together of such simple compounds as carbon dioxide, water and ammonium compounds to produce a variety of active and inactive organic molecules. The second part of the cycle is the utilization of the stored energy by both the photosynthetic and non-photosynthetic organisms through a series of complicated steps; this is done either to synthesize essential molecules or macromolecules, or to produce energy for the processes of life. But no matter what path this utilization follows, the final result is the gradual release of the stored energy and the return of the simple compounds to their natural forms. Then most of this released energy finds its way into the energy cycle on the surface of Earth and is eventually radiated back into space. The outcome of these activities is an orderly flow of energy from the Sun through a relatively thin layer of organic material covering the Earth; thus the total amount of energy held in storage by the living and their by-products remains more or less constant. The reason for this constancy, however, is not the inadequate quantities of radiation, but rather the restricted space for the living and, depending on the location, the limited supply of the essential elements and compounds.

The total dependency of life on solar energy and the manner of movement of this energy through the system indicates that the utilization of the Sun's radiation has been an integral part of the evolution of life right from its beginning. This makes the origin of life closely interwoven with the evolution of the Earth and the Sun. In other words, *the foundations of life were laid during the Earth's origin.* We can even narrow down the period in Earth's evolution when this took place. We know that none of the simple but fundamental organic compounds associated with life can withstand the high temperatures that existed during the Earth's molten era. Therefore, most probably the prebiotic compounds appeared after the surface of Earth had become solidified but before most of the water vapors in the atmosphere had condensed.

Self-replication and Origin of Life

In spite of the fact that the radiation from the Sun is the power behind life, that energy would not have been of value had not the development of the capability for reproduction by the emerging units already taken place. In other words, the process of primitive self-replication came into being before the development of metabolic mechanisms. This could have happened if the emerging self-replicating units were continually supplied with their energy needs through non-biologic and non-specific mechanisms.

Another reason for the belief that self-replication has had precedence over metabolic activities is that this phenomenon is not and has never been the monopoly of the living. Self-replication is as old as *matter* itself. In the creation of matter the subatomic particles have continually catalyzed the creation of their own kind; in the formation of stars, star-aggregates have accelerated the origin of other stars, and the galaxies have laid the foundations for the development of other galaxies. The original self-producing molecular aggregates on Earth were following the course of a very natural phenomenon. In the beginning, the replication of the living units most probably followed the same crude pattern that has been the mark of the inanimate. The refinement came later when, instead of purely physical and chemical forces and energies, biological compounds in the form of complex organic molecules became the backbone of the process, resulting in the formation of delicate, adaptable and versatile systems.

Once self-replication appeared among the molecular aggregates, it continued automatically without regard for consequences. This means that even in those times, competition among the units was a prevailing phenomenon. The more efficient systems consumed the most beneficial compounds and deprived the less efficient units from receiving their material needs. In the beginning, the rate of reproduction was very slow and the material supply relatively plentiful. Consequently, there was very little competition. But as the rate of reproduction increased and the supplies were consumed at faster rates, the more efficient and hence the stronger evolving primitive units caused the elimination of the weaker ones and in many cases utilized their contents. This was the beginning of the phenomenon of natural selection, the automatic consequence of competition between varieties of living units for the attainment and consumption of limited supplies of usable material. Those that could adapt to the harshness of the environmental conditions survived and those that could not were eliminated. It is interesting to note that, similar to the replication phenomenon, the struggle for survival has never been restricted to the living. The same fundamentals have been in operation among stars that have had to share their atmospheric gases and among galaxies competing for the annexation of the available stars.

Fundamental Requirements for the Origin and Evolution of Life

One significant fact about any major evolutionary trend is that the factors initiating the evolution remain with it through its entire course. The characters or identities of many of these factors may alter, some to a great extent, but enough of the fundamentals remain behind to provide clues to the probable conditions which caused the evolutionary origin. Accordingly, the present-day requirements for the continuation of life on Earth are of value in reconstructing the conditions needed for its origin. Without them life has no chance; with them it may or may not originate and evolve. And if it does, the success of the evolution is still a doubtful proposition.

A planet may be considered as having the possibility for bearing life if it meets the following correlated conditions:

1. The mass of the planet should be large enough to prevent the escape of low atomic weight elements and compounds, especially at the relatively high temperatures associated with the initial stages of its evolution. At this point attention is drawn to Chapter V (The Evolution of the Solar System) in which it was stipulated that the gravitational attraction of a planet holds back considerable amounts of antagonistic gases in the form of a shell above its homogeneous atmosphere. It was further implied that the radiation zone between the antagonistic shell and the atmosphere below prevents the escape of lightweight elements and compounds. Now, as long as the antagonistic cover surrounds the planet, the essential light atoms and molecules cannot escape. Only after the depletion of the antagonistic gases and the resulting cooling of the planet does its mass become a determining factor in the retention of such gases as methane, water vapors, ammonia, carbon dioxide, carbon monoxide, etc.

2. The conditions on the surface of the planet should be such that after it has cooled down and its state of equilibrium established, the temperature variations on a large area on its surface neither exceed the temperature of boiling water nor drop too far below the freezing temperature of water. Also, at certain regions, preferably around the equator, the temperatures must remain mild for the water to stay continuously in the liquid phase for billions of years.

3. The maintenance of a mild and even temperature requires the presence of an adequate atmosphere. Such an atmosphere has the added function of protecting the evolving units from the harmful effects of excessive radiation and cosmic rays.

4. The presence of large quantities of water is an absolute necessity for the origin and successful evolution of life. Although it may be theorized that other forms of life can originate within liquids other than water, all the available evidence shows that the primary functions of life cannot be performed without the presence of appreciable amounts of water in its aqueous phase. In addition to the chemical and the physical properties of water that are needed for the

processes of life, large bodies of water are required for the purpose of buffering the variations in the surface temperature of the planet, diluting the waste products and in general holding the very delicate threads of life together.

5. The amount of radiation that a planet receives from its central star determines whether or not it has the possibility of possessing mild environmental conditions. Therefore, if a planet is to bear life, it should be at a proper distance from its star.

6. The planet should possess a correct ratio of heavier elements. This necessitates a considerably long period of contact between the antagonistic gases and the evolving planet. One of the functions of such heavy elements is to produce the solid framework that can hold the planet intact at the mild temperatures needed for the perpetuation of life.

7. One of the prime requisites for a planet to become life-bearing is that it should rotate at a fairly rapid rate along an axis approximately perpendicular to the orbital plane. This type of motion assists in the establishment of three of the needed conditions for the origin and the continuation of life: (1) a rapid rate of rotation causes an even distribution of energy received from the star, thus greatly aiding in the maintenance of a mild temperature with tolerable fluctuations; (2) it is instrumental in producing a fairly strong magnetic field which satisfactorily controls the entry of damaging cosmic particles; and (3) it provides a cyclical system of days and nights which stimulates the development of cyclical systems among some of the evolving units.

With regard to the last section of the above paragraph, by creating conditions for the primitive forms of life to receive energy from the Sun intermittently, the rotation of the Earth has most probably played an important role in the evolution of metabolic cyclical systems among the pre-cellular molecular aggregates. The regular interruptions in the arrival of the solar radiation have most probably been instrumental in making the early molecular aggregates adapt their mechanisms to the cyclical pattern of energy arrivals—that is, they would have accumulated energy during the day and discharged it during the night. In other words, ADP-type molecules that could store energy when they were subjected to radiation changed to ATP-type compounds during the day, and then at night these high-energy molecules transferred their energies to other molecules and transformed back to ADP-type molecules. We shall elaborate on these steps later on, but at this time we should like to point out the similarity of this scheme with the general metabolic activity of plants. In spite of their highly sophisticated metabolic systems, the charging and discharging of energy in plants follows the day-and-night pattern: during the days the energy of solar radiation is stored in certain molecules and during the nights part of this stored energy is used for respiration.

Inter-relationships Between the Origins of Life and Earth

1. Physical Conditions on Earth During the Initial Stages of Its Origin

As long as the antagonistic shell was completely covering the homogeneous atmosphere of the Earth, the planet remained hot and gaseous; but when the depletion began to show its effects, the temperatures dropped slowly and the Earth's surface gradually solidified. A very rough estimate for this period would be between one hundred and two hundred million years. However, it took a much longer time for the complete disappearance of *antiH* gases; some lingered on for perhaps another one to two hundred million years, mixed with the upper atmospheric gases in the manner of the contemporary radiation-emitting nebulae.

With the gradual disappearance of the antagonistic gases the temperatures dropped at a steady pace. At between 250° C and 350° C (the condensing temperatures of large quantities of water vapors under very high pressures) the water vapors began to condense. The conditions were then ready for the formation of the simple prebiotic compounds. The first of these compounds probably appeared when temperatures were between 300° C and 400° C, and by the time temperatures had dropped to about 100° C, sufficient amounts had come into existence to provide the basic compounds for the formation of the most primitive self-replicating macromolecules. The details of these processes will be presented shortly.

It is evident that these proposals characterize an atmosphere saturated with water vapors which condensed very slowly. These are not the conditions that would permit the synthesis of biologically oriented compounds according to some of the theoretical pathways found in the literature. Those chemical schemes that require the absence of water for the formation of prebiotic compounds would not fit in with our proposals, but the ones that can occur in the presence of water have been of great help in planning the pathways to be presented in the following pages. The most important of these are the formation of amino acids from ammonia and methane, and the condensation of formaldehyde to produce monosaccharides.

Under the violent and varying conditions of the prebiotic times, all types of reactions could have taken place and many compounds produced through a variety of pathways. Thus, the reconstruction of the reactions that took part in the formation of prebiotic compounds is a highly speculative undertaking. With the available knowledge all one can do is to propose chemical schemes and say—"the probabilities favored such and such reactions." In fact, we must always keep in mind that conditions on the evolving Earth in no way resembled those under which controlled experiments are performed in laboratories. The cooling planet had an inhospitable and changing environment. The continual annihilation of matter in the upper atmosphere and the wholesale condensation of water vapors over the hot and eruptive crust of the planet constantly disturbed the balances between the interacting compounds. These are not conditions conducive either to the formation of large quantities of organic matter or their accumulation in high concentrations in the primordial aqueous basins from where life could begin. Therefore, any proposed hypothesis for the origin of life should be based on the continual production of the prebiotic compounds and their steady inflow into these basins before and during the origin of the self-replicating molecular aggregates.

We have already mentioned that most of the fundamental factors taking part in the origin of any evolutionary trend remain with the evolution throughout its life; most of them usually alter, some considerably, but generally the identities remain recognizable. With this in mind, the current methods of approach to the problem of the origin of life become questionable. The important drawback in the current theories, we believe, is the acceptance of the principle that the prebiotic compounds were of the same chemical compositions as the contemporary biological compounds performing similar functions. It cannot be denied that the presently known compounds are related fairly closely to the prebiotic ones, but it would be futile to attempt to reconstruct the composition of the primordial self-replicating macromolecules with them. Such an undertaking would be tantamount to attempting to rebuild the Wright's original airplane (assuming we did not have any knowledge of its structural details) with parts that are being used in the manufacture of jet planes. In order to acquire precise knowledge of what occurred in the distant past, we must learn about the real identities and the structural details of the prebiotic compounds, and to accomplish this we have to reenact as closely as possible the events of the primordial periods in well-conceived and patiently conducted experiments. We cannot simulate the $+matter\text{-}antiH$ reactions, but since we have a fairly good idea of the manner of energy production and the nature of the resulting subatomic particles, it would be possible to set up experiments that would reasonably represent the conditions of the prebiotic times. These should give us the necessary information regarding the real nature of the compounds involved in the origin of life. We may even discover some essential compounds unsuspected of having any association with life or its origin. (In this respect it would be interesting to determine the exact molecular structures of the organic compounds below the surfaces of the dried up water basins on Mars.)

It may be argued that the generation of electromagnetic energies and the ejection of subatomic particles at close distances as well as the cosmic particles originating from the evolving solar system would have had deleterious effects on a considerable number of delicate organic compounds needed for the origin and perpetuation of life. However, it should be remembered that in these discussions our prime target is not the possibility of the occurrence of certain reactions; our main consideration here is the discovery of the manner of development of a trend resulting from the interactions between constructive and destructive forces acting on each other. Obviously the balance came out in favor of the origin of life.

2. *The Influence of the Antagonistic Shell on the Formation of Prebiotic Compounds*

The antagonistic gases that remained with the Earth after the completion of the recession of the *antiH* clouds performed five important functions: (1) during the period when the temperatures were high, the radiation zone prevented most of the lightweight molecules (not hydrogen and helium) from escaping; (2) by continuously generating energy, they slowed down and stabilized the rate of the Earth's cooling; (3) they caused the continual production of compounds and free radicals similar to the ones observed in comets and nebulae (the same types of reactions were taking place); (4) they released subatomic particles and radiation energies at a steady rate for a long period of time; and (5) by remaining in the upper atmosphere long after most of the water vapors had condensed (probably between one to two hundred million years), they provided a steady supply of mild energy, electrons and low-energy subatomic particles; all of these were needed for the continuation of the replication processes of macromolecules in some of the water basins on the Earth.

With regard to the contents of the above paragraph we are proposing the following:

1. The interactions between the free radicals resulting from reactions mentioned in the third section, aided by continual inflow of energy from above, brought forth the synthesis of a whole series of aliphatic and aromatic hydrocarbons as well as a variety of related simple nitrogenous compounds.

2. Some of the hydrocarbons contained one or more C^{14} isotopes which on transmutation were turned into N^{14} atoms and became converted to such nitrogenous molecules as purines, pyramidines, pyrolle rings and other nitrogen-containing compounds.

3. The non-specific linkage of molecules through the de-hydrolysis process could have come about by the ionization of the hydrogen atoms or the oxygen portion of the hydroxyl radicals; this ionization was caused by radiation and possibly by low-energy electrons produced in the upper atmosphere.

The formation of hydrocarbons from free radicals and the synthesis of the related simple nitrogenous compounds through the reactions between the hydrocarbons, CN and NH_3 radicals, is well known and needs no comments. In the following paragraphs we shall describe the probable manner of formation of some of the complex nitrogenous compounds by the transmutation process and two possible ways nonspecific dehydrolysis could cause the joining of two adjacent molecules to produce macromolecules.

The study of the synthesis of the nitrogen-containing compounds from the naturally occurring hydrocarbons or their derivatives by the process of transmutation of C^{14} to N^{14} is an overlooked branch of organic chemistry. The oversight is understandable because this manner of synthesis has to rely on the probability that the C^{14} isotope, with a half life of about five thousand years, would transform to N^{14} at a specified location in the molecule; this is an impractical way of synthesizing compounds. Yet the possibility exists that this type of transformation has played a significant part in the synthesis of purines, pyrimidines and other nitrogen-containing molecules prior to and during the origin of life. Most probably the C^{14} atoms that participated in this type of synthesis were formed during the +$matter$-$antiH$ $reactions$ and before the formation of the hydrocarbon molecules. It does not seem likely that many of the C^{14} atoms produced afterward from nitrogen sources in the atmosphere (as is occurring at present) could have found their way into the hydrocarbon molecules. Also it should be noted that during the initial stages of evolution, C^{14} was much more plentiful than at present.

The present state of our knowledge does not permit the formulation of chemical schemes giving the possible manner of the formation of the more complex nitrogen-containing compounds by transmutation of C^{14} to N^{14} . However, simple examples can be given as to how some of these compounds could have become synthesized. For instance, it is theoretically possible to produce pyrimidine from benzene:

Similarly, a pyrrole ring, the building blocks of porphyrines, can probably be produced from 1,3-cyclopentadiene:

The same principle may be applied to the aliphatic series in the formation of amines, amides and certain amino acids. For example, if the carbon atom of a -CH$_3$ radical attached to an aliphatic chain were a C^{14} isotope, the radical would be transformed to an amino group after the completion of the transmutation of the C^{14} atom to a nitrogen atom, producing a compound with an amino radical. It might be mentioned here that this type of transformation could have also produced significant mutations of genes during the post-cellular era of evolution of life.

In the biologic world the process of dehydrolysis is the most utilized step for the joining of molecules to produce macromolecules. In the contemporary systems, enzymes usually perform this function by catalyzing the removal of one molecule of water from the ends of two specific molecules—an OH from one end and an H from the other. But since in the prebiotic times there were neither enzymes nor other catalysts that constantly performed this function, the removal of water from the two adjoining molecules would have had to occur by direct, non-specific means. For this to happen, the OH and the H endpieces had to become activated by physical agents after the OH radical of one molecule had become lined up next to the H atom of the second molecule (the probable steps for the lining up of molecules will be discussed presently). Once the lining up had taken place, then non-specific dehydrolysis would bring about their unification. The removal of water from the two molecules could have taken place through one or both of the following processes: first, the ionization of hydrogen or oxygen atoms (possibly by radiation or the low-energy subatomic particles); and second, by direct chemical reactions, with the prevailing high temperatures speeding the process.

The First Evolutionary Stage — The Origin

Before we proceed with the description of our proposed hypotheses about the first two evolutionary stages of life, we should once again contemplate whether the origin of life was the result of accidental encounters of the right organic compounds in the right concentrations at the right time and in the right place, or whether it was an inevitable series of events that resulted from the conditions created by the evolution of the Earth. This is a fundamental decision because, if the origin of life was an accident resulting from a large number of coincidental events (each of which was subject to the uncertainties of the law of chance), then life is indeed a very rare phenomenon in the universe. But if the origin required the fulfillment of certain conditions with only some assistance from the law of probability, then life could originate on any planet that could sustain it. This would make life a natural evolutionary trend that comes into existence within the framework of the evolution of matter. We believe in the second alternative. However, we do not share the opinions of some of the enthusiasts about the omnipresence of life in our Galaxy and the universe. The reason for this view is that for life to evolve on other planets, the delicately interwoven reactions responsible for the origin of life have to be repeated in almost the same way as occurred on Earth. The great weakness of the evolution, particularly in its first stage, is its susceptibility to complete disruption by a variety of cosmological, interstellar or interplanetary mishaps. If life has begun and its progress is then brought to a total halt, it has very little chance for a new start, even if the environment becomes favorable again. Life can originate only during a specific period in the course of the evolution of a planet, and if that opportunity passes by without a successful outcome, the probabilities do not favor a repeat performance.

Aggregation of Prebiotic Compounds

The first step toward the formation of life on Earth was the orderly aggregation of the prebiotic compounds in fresh waters containing adequate amounts of soluble phosphates. The depth of the water was of paramount importance, as waters too deep would have obstructed the passage of the needed energies and waters too shallow would have permitted the detrimental effects of the high-energy radiation and high-velocity subatomic particles to disrupt the sequence of evolutionary steps. Therefore, the most likely place for the origin of life was a comparatively large freshwater lake fed by slowly flowing phosphate-containing streams. Since a number of lakes in different locations on Earth could meet these conditions, most likely the initial processes took place in a number of lakes, but the chances are that life emerged from only one of them.

During the early stages of life, the Earth's atmosphere and the "life-supporting" lake had the following properties:

1. The temperatures of both the lake and the atmosphere were over 100° C; the very high moisture in the atmosphere as well as the +*matter-antiH reactions* in the upper atmosphere prevented the temperatures from dropping at a rapid rate.

2. The principal components of the atmosphere were water vapor, ammonia, methane, formaldehyde, carbon dioxide, carbon monoxide, hydrogen cyanide, nitrogen, hydrogen,

some oxygen, innert gases, and a mixture of low-boiling aliphatic and aromatic hydrocarbons and their water-soluble derivatives.

3. The pH of water in the lake was either neutral, or more probably slightly alkaline, buffered by the phosphate salts. This condition protected the newly formed macromolecules from becoming easily hydrolyzed at the prevailing temperatures.

4. No delicate state of equilibrium had a chance of lasting for long in the atmosphere. The intense radiation and the incoming high-energy particles were continually causing the formation and destruction of organic compounds. Many of the soluble forms became dissolved in the condensing waters and were carried into the lake.

5. Some ozone could have been produced in the atmosphere, the oxygen coming from the photolysis of water vapors. If this was the case, then the ozone protective layer was formed early in the life of the Earth.

6. The water in the lake protected the dissolved compounds and the newly formed macromolecules against destruction by injurious agents. Not only did the surface water layer prevent the intense radiation from reaching the deeper levels, but it also slowed down the rates of movements of the high-energy particles. Additional protection also came from the suspended inorganic colloidal particles which limited the intense activities to the surface areas.

7. The continual inflow of waters brought all types of compounds into the lake, but the outflow kept the concentrations of salts, organic compounds and the colloidal particles within a moderate range.

8. Some of the organic compounds brought into the lake were monosaccharides (probably produced by the condensation of atmospheric formaldehyde). Under the influence of high temperatures, many of these simple sugars became phosphorylated by reacting with dissolved phosphate radicals.

The picture presented here depicts a haphazard formation of a variety of simple organic compounds in the atmosphere, with some finding their way into the lake; but most of these molecules were not suitable for the origin of life. The prebiotic compounds were highly diluted, and the only way they could come together was through their selective adsorption on certain colloidal particles floating in the water.

Formation of Primordial Helices

In this book we are proceeding with the assumption that the self-replicating macromolecules were the first to appear on the scene; then as their evolution progressed, they formed polypeptides and proteins in the image of their own structural patterns. It is also believed that competition between the units for the available material was responsible for these additions to the already functioning self-replication process. This may be interpreted as the primitive form of the phenomenon of natural selection.

In the theorteical reconstruction of the origin of the self-replicating units, a considerable amount of flexibility can be given to the chemical structures of the prebiotic purines and pyrimidines; still it is not possible to replace the skeletal structure composed of either ribose-to-ribose, or deoxyribose-to-deoxyribose, molecules connected together through phosphate radicals. The reason for this difficulty lies in their structural forms. Connected together by phosphate radicals, they produce the only possible structure that can produce the helices with the capability of self-replication. Thus we can be fairly certain of the presence of these well-known pentoses during the prebiotic times. However, it would be unreasonable to assume these to be the only carbohydrates formed. Probably a variety of monosaccharides came into existence either by the condensation of formaldehyde molecules, the hydroxylation of the aliphatic chains or by some other as yet unknown pathway. Many of them probably made phosphate linkages, but only the ribose or deoxyribose phosphate molecules could produce the supporting frameworks for the self-replicating macromolecules.

The first step toward the formation of the self-replicating macromolecules was the formation of "ribose-to-ribose" or, more likely, "deoxyribose-to-deoxyribose" skeletal structures. Two conditions were needed for the fulfillment of this step.

First, adequate quantities of colloidal particles that could adsorb these monosaccharide phosphate molecules non-specifically should have been present in the lake. Because of their pronounced charges, the phosphorylated pentoses could have been easily adsorbed on the right particles through the attraction between one of the phosphate's acidic groups and the adsorbent. Second, after the non-specific attachment to a particle had taken place, the new combination would turn the adsorbing particle into a vehicle for the specific adsorption of a molecule identical to the first one. This could have happened if the adsorbed molecule provided a second adsorbing point at a specific position relative to the first adsorbing site.

According to this general scheme, once an L-deoxyribose phosphate molecule is adsorbed, the second molecule has to be an L-deoxyribose phosphate molecule or it will not be adsorbed in the proper position next to the first molecule. In the coming paragraphs we shall describe how various helices were formed. Here, we draw attention to the impracticability of the formation of helices of monosaccharide phosphates composed of other than ribose or deoxyribose phosphate molecules; their molecular configurations do not permit the synthesis of the ideal helically shaped frameworks.

In the following discussions we shall use only deoxyribose for our schemes and examples. This is with the understanding that the same could have happened in the formation of ribose helices.

The formation of helical chains from identical pentoses with the same optical rotation could have occurred if, prior to their chemical unification, the phosphorylated deoxyribose molecules had been selectively adsorbed one at a time in the correct position on the adsorbing particle. This could have happened if the first molecule A (Figure 54-step I) had one of its acidic groups adsorbed to the surface of the particle while its second acidic group remained free to produce a point of anchorage for the positioning of a second identical molecule. The second adsorbed molecule B (Figure 54-step II) would then be in a specific position with respect to the first one in such a way that its OH radical of the 3' carbon atom would be exactly opposite the free acidic group of the phosphate radical of the first molecule A.

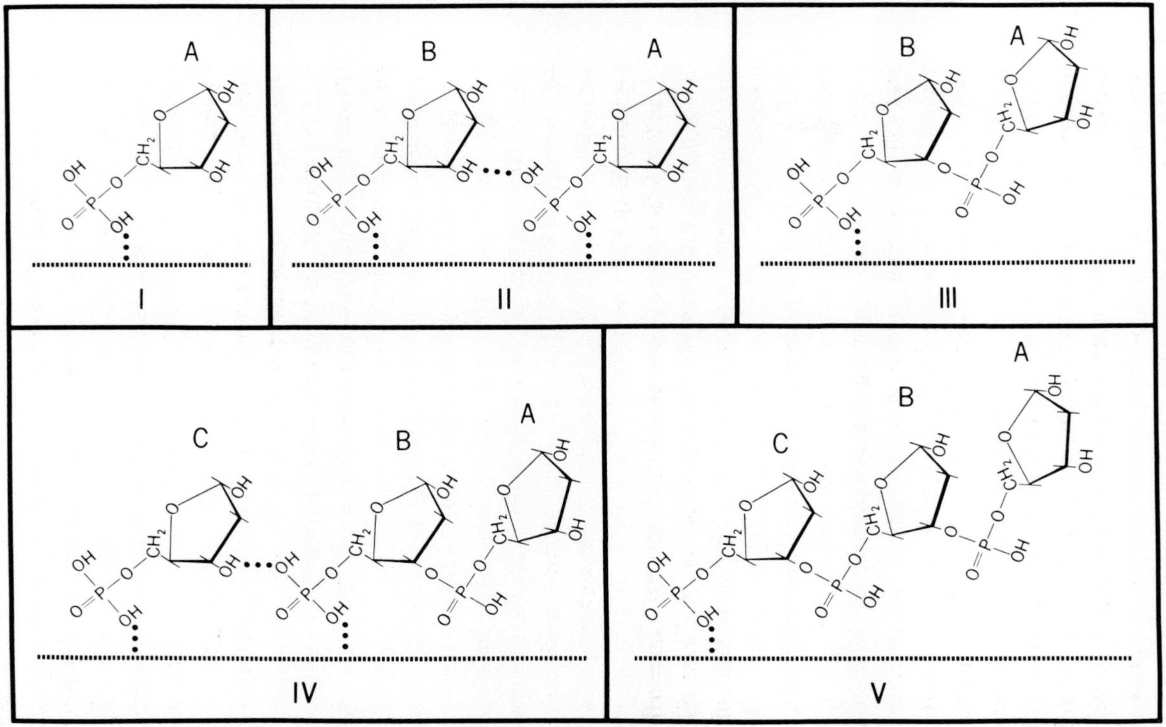

Figure 54. SUGGESTED PATTERN OF ADSORPTION AND CHEMICAL UNIFICATION OF THREE DEOXYRIBOSE PHOSPHATE MOLECULES ON THE SURFACE OF AN ADSORBING PARTICLE. The deoxyribose molecules are shown as two-dimensional structures. However, there is a twist in the molecule which, after chain formation, causes the macromolecule to be in the form of a helix.

This would produce a hydrogen bond between the two active groups. The second molecule would then remain in this position with respect to the first one until non-specific dehydrolysis would cause their chemical unification. The chemical combination would then bring about the twisting of the molecule A which would result in the breakage of its hydrogen bond with the adsorbing particle (Figure 54-step III).

The steps IV and V of Figure 54 shows the manner in which a third molecule is added to the chain. As may be seen in the diagram, it is always the last adsorbed deoxyribose phosphate molecule that remains attached to the particle. The rest are free to assume their helical configuration. As a consequence of this manner of chain-building, the adsorbing particle cannot hold to the growing helix except through a hydrogen bond with the latest annexed molecule.

Each deoxyribose molecule in the growing helix had a potential site on its 1' carbon atom for adsorbing a purine or a pyrimidine molecule. After their adsorptions these base molecules would then become unified chemically with the skeletal frame by non-specific dehydrolysis. The whole process may be pictured as a gradually growing helix appended to the adsorbing particle at its growing end. Each time a new deoxyribose molecule was added to the chain, the new monomer would become the anchoring agent for the whole helix. Then when the opportunity arose, a purine or a pyrimidine molecule would become attached to the 1' carbon atom of the pentoses in the helix. This process may be considered as the first step toward the synthesis of primitive nucleic acid macromolecules.

In due time the growing macromolecule would break away from the adsorbing particle at a weak point, probably at the relatively weak hydrogen bond existing between the deoxyribose molecule attached to the particle and the second monomer, prior to their chemical combination. The end piece remaining on the particle would then begin to grow again, with the system having the potential for producing a number of nucleic acid chains. It is evident that the primitive nucleic acids thus produced could neither have been of the same lengths nor have had a definite pattern for the arrangement of their nitrogenous bases on the helix. The frameworks were composed exclusively of either D or L deoxyribose helices, but there were no controls over the breakage points nor over the types of purines or pyrimidines that became attached to the central skeleton. It is also clear that the lengths of the longest of these macromolecules were extremely short as compared to the lengths of strands in contemporary cells.

Formation of Primordial Double Helices

The next major step in the evolution of macromolecules was the formation of the complementary strands with the newly formed primitive nucleic acid chains. For that, a continuous supply of nucleotides was required. A number of pathways may be devised for the formation of these units, but the most plausible one seems to be the degradation of the deformed nucleic acid macromolecules. It should be remembered that the temperature of the lake in question was close to $100°$ C, and the possibility existed for some of the nucleic acid strands, especially the degenerated ones, to become hydrolyzed.

The process of the formation of the double helices consisted of the adsorption of series of complementary deoxynucleotides along the lengths of primary strands. Chemical unification of the adsorbed molecules by non-specific dehydrolysis would then cause the formation of the complementary strands. For example, an adenine-type base on an original strand would attract a deoxynucleotide molecule containing a thymine-type base, and a guanine-type base would form a hydrogen bond with a cytosine-type base. A series of such adsorptions would produce a reciprocally related strand of complementary deoxynucleotides along the length of the helix, positioned with respect to each other in such a way that all the OH groups on the 3′ carbon atoms of their deoxyribose molecules would be situated next to the free acidic groups of the phosphate radicals of the neighboring units. Then gradually non-specific dehydrolysis would cause their chemical combination and produce the complementary strand of the double helix.

The whole question of the origin of life revolves around the capability of macromolecules to self-replicate. For this, the double helices would have had to become unwound at frequent intervals. But the double helices were fairly stable structures, and this makes it difficult even to conjecture why the two strands should have separated regularly. A good guess would be that some non-specific mechanism, such as the ionization caused by sudden increases in the rates of radiation from the juvenile Sun, would have caused the breaking up of the hydrogen bonds between the two units—but this is not totally a satisfactory explanation. The answer may eventually be derived from the complete elucidation of the unwinding mechanisms of the present-day helices. Although these strands are much longer and their self-replication systems much more complex, the basic mechanism for their unwinding is probably quite similar to that of the primitive strands. In any case, irrespective of the causes, we can be certain that the regular unwindings of the double helices took place during the origin of life.

Formation of Ribose Phosphate Helices

While these developments were taking place, the ribose phosphate molecules were also contributing their share toward establishing the essential processes of life. It has already been mentioned that deoxyribose molecules were not the only type of pentose produced during the primordial times and that probably the ribose molecules became just as abundant as their deoxy-relatives. This means there was just as much chance to produce helical structures composed of ribose phosphate monomers as of deoxyribose molecules. It is believed this group of compounds took part in the processes of the origin of life through two major steps. The first one consisted of the ribose phosphate molecules going through the same physical and chemical reactions as described for the formation of deoxyribose helices; but since the ribose macromolecules had no possibility of self-replication, they remained in solution until hydrolysis caused their breakdown to nucleotides.

In the second stage these nucleotides formed RNA-type strands in the imprints of the DNA-type macromolecules; the mechanism of their formation was the same as the one described for the formation of the complementary strands. The main difference between the present and the primordial methods for the formation of the RNA-type strands seems to be in the manner the phosphate and hydroxyl radicals became combined. In early times, the chemical unification went through successive steps of specific adsorption followed by the joining of the molecule by non-specific dehydrolysis. In contemporary processes, specific enzymatic activities control all the reactions.

The Second Evolutionary Stage — The Adjustments

It would be safe to assume that some of the primordial RNA-type macromolecules that possessed the right combination of nitrogenous bases acted as templates for the formation of the original polypeptide chains. However, the mechanisms for the synthesis of these primordial polypeptides are obscure, and considerable amounts of experiments are needed to find the true pathways for the synthesis of this type of primitive molecular arrangement without the participation of enzymes. Yet, no matter what the pathway, the fact is that present-day ways of synthesizing polypeptides and proteins have been evolved from similar but much simpler systems that came into being during the origin of life. To summarize then, the main process consisted of the DNA-type patterns being imprinted on the RNA-type strands which, if possessed of the right combination of bases, would line up the amino acids in orderly sequences; non-specific dehydrolysis would then join them together to produce related polypeptide chains.

Formation of Molecular Aggregates

The appearance of polypeptide chains, a complicated step requiring a considerable number of accurately positioned active sites on the helices, narrowed down the chances of survival of a large proportion of the population of the self-replicating macromolecules. Among the countless varieties that existed, only a small number could have developed the power to synthesize polypeptide chains. By itself, the synthesis of the polypeptide chains was not a determining factor in the battles of survival that were to happen, but since it was the forerunner of the capability for the synthesis of proteins and hence enzymes, those macromolecules that became capable of making and using them to advantage became better prepared for the approaching struggles. The advent of the polypeptide synthesis, therefore, was the beginning of the era during which the number of lineages became considerably reduced while the struggles for survival became intensified.

Many of the units that succeeded in synthesizing the RNA-type strands and polypeptide chains became physically attached to the newly developed compounds to form molecular aggregates. Although most of them could replicate for a few generations, only a small fraction of them contained the components that could function in a coordinated manner and hence multiply regularly. Structurally, each aggregate was composed of a DNA-type double helix surrounded by RNA-type strands and related polypeptide chains. Since all the components had come into being in the pattern set by the sequences of the DNA's nitrogenous bases, they held together firmly as a unit.

In most of these aggregates, the polypeptide chains could not participate in the functions needed for self-replication processes. Hence they became useless structures that were more detrimental than helpful in the activities of the aggregates; these units could not keep up with the competition and gradually perished. There were some units, though, in which each component (including the polypeptide chains) contributed its share to the maintenance and functions of the aggregate; these were the ones that eventually gained dominance over the hierarchy of the entire self-replicating population. *A new phenomenon had thus made its appearance: the differentiation of functions among the components of a unit with the aim of benefiting the whole.* This phenomenon evolved with the evolution of life. It can be seen in all facets of present-day life—all the way from the harmonious and coordinated activities of the components of individual cells to the inter-related contributions by the differentiated cells of the multicellular organisms. Even the social behavior among animal tribes and human societies is not exempt from the influences of this deeply rooted heritage.

Appearance of Primitive Polypeptide Enzymes

In those early days, the fittest molecular aggregates were the few types that came into the possession of the polypeptide chains with the capability of non-specific hydrolysis of nucleic acid molecules and polypeptide chains of the less fit units. This mode of obtaining the scarce material for the preservation of the individual unit and the continuation of the lineage must have been dependent on encounters based on probability: either the aggregate came directly in contact with its "food," or could not break it down and assimilate the contents. The important aspect of this development, however, was that a new factor had been introduced for achieving dominance—that is, the use of biologic weapons by means of which the poorly defended could be subdued and their material possessed. Similar to the phenomenon mentioned in the previous paragraph, this one also evolved with the evolution of life.

By this time, which was approximately between twenty and seventy million years after the first biotic compounds had made their appearance, most of the water vapors in the atmosphere had condensed and a fair quantity of the liquid had passed through the birthplace of life. The lake became filled beyond its capacity and the overflow carried with it some of the self-replicating macromolecules. However, it is doubtful if at that time many of the lower basins, especially the oceans, had the suitable environment for their proliferation. An important factor in the prevention of effective reproduction was the lack of an ample supply of soluble phosphate salts. But life continued to evolve in the original lake and the reservoirs with approximately similar environmental qualities. The temperatures at this time must have been below 100° C and have continued to drop at a slow rate. The concentration of the antagonistic atoms in the upper atmosphere had also been considerably reduced, and their influences on the formation of new macromolecules were gradually diminishing. The molecular aggregates either had to develop the mechanisms for the utilization of the Sun's energy or cease to replicate. Approximately fifteen to fifty million years were left for them to accomplish the deed.

The same factors which were instrumental in bringing about the formation of the hydrolyzing polypeptides were also responsible for bringing into existence their opposite numbers, the dehydrolytic polypeptides. We know that there is little difference between the structural configurations of the active sites of hydrolyzing and dehydrolytic enzymes. Therefore we can safely assume that the enzyme-like polypeptides that could cause the removal of water from two adjoining compounds were structurally very close to those that caused hydrolysis. We can also say that the dehydrolytic polypeptides came into being some time after the hydrolyzing ones. Contrary to hydrolytic actions, the removal of water required the intake of energy via the AMP, ADP and ATP molecules. These could have been produced by the hydrolysis of RNA-type macromolecules. As mentioned previously, the initial path for the transfer of energy was probably through ADP-type and ATP-type systems, with the charging taking place during the day and the discharging during the night. By this time the continuous energy coming from the antagonistic shell had weakened considerably and the rotation of Earth was causing variations in the amounts of energy received, thus producing semblances of days and nights. Probably the units of the advanced aggregates contained many ADP-type and ATP-type molecules. As a result, considerable amounts of energy could move through an aggregate, even though the charging of the molecules took place only once a day.

The introduction of the dehydrolytic polypeptides and the energy transferring nucleotides into certain groups of aggregates gave them the capability of performing two vital functions simultaneously: they could break down the less advanced macromolecules they encountered, and then utilize the products of the breakdowns. The hydrolysis of the less fit aggregates produced a variety of deoxynucleotides, ribonucleotides and amino acids which the dehydrolytic polypeptides could use to increase the lengths of their DNA-type helices, RNA-type strands and polypeptide chains. However, although many macromolecules became longer, only a limited number of DNA-type helices received the right combination of the sequence of the nitrogenous bases to become instrumental in producing the more advanced types of enzyme-like molecules. Their offensive and defensive weapons were increasing in number and complexity. Consequently the potential units of life were becoming larger and more efficient; they were moving toward self-sufficiency.

Appearance of Primitive Photosynthesis

Up to this time the origin of life had progressed at a fairly rapid pace, but with the continual reduction in the number of *antiH* atoms in the upper atmosphere and the slowing down of the rate of synthesis of the vital compounds, this state of affairs was fated to change. The evolution had entered a very crucial stage. Unless some of the aggregates developed the needed mechanisms for the synthesis of the disappearing essential compounds, the whole evolutionary movement would have come to a halt. There was a considerable amount of waste in the transfer of the material of one aggregate to others; hence, while self-replication was continuing, the supplies were getting scarcer.

Salvation came from the process of photophosphorylation that was already in operation in conjunction with the dehydrolytic polypeptides. Many of the more advanced aggregates already possessed the ADP-type and ATP-type hydrolyzing and dehydrolytic systems which allowed them a limited utilization of the solar energy. These simple processes, though, could not produce the more complex molecules for the continuation of the evolution. The mechanisms for the utilization of energy and the use of the energy for the synthesis of the essential compounds had to improve greatly. This improvement could come through the incorporation of the photosensitive pigments (such as complex molecules with pyrolle rings, already synthesized and available in the lake) into the energy-transfer mechanism. Obviously, some of the aggregates succeeded in developing such a system, and as a result primitive photosynthesis made its appearance.

Disappearance of Aggregates with L-Pentose Helices

In the early part of this discussion it was indicated that both L and D pentoses could participate in the formation of the monosaccharide phosphate helices, but since the adsorption of molecules on the particle was a specific process, the resulting framework was composed either of L or D pentoses exclusively. Therefore, an aggregate could belong to either one of the two groups: the first group included the aggregates with DNA-type and RNA-type strands, which contained D-pentoses and polypeptide chains composed of L-alpha amino acids; the second group consisted of aggregates with DNA-type and RNA-type strands that contained L-pentoses and polypeptide chains composed of D-alpha amino acids. These statements are based on the belief that during the synthesis of polypeptides, the direction of the curvature of the helices in relation to the spatial arrangement of their nitrogenous bases necessitated an orderly adsorption of either L-alpha amino acids to the D-pentose frameworks or D-alpha amino acids to the L-pentose skeletal structures. Chances are that about equal numbers of each group came into existence.

It is believed that the appearance of photosynthesis among a specific group of aggregates caused the elimination of the "L-pentose-D-alpha amino acids group" from the evolutionary scene. The cause of the disappearance was most probably this group's inability to develop efficient photosynthetic processes in time. The units which became proficient in utilizing the solar energy were not only capable of proliferation at an accelerated rate, but were also the sources of the needed compounds for the more advanced non-photosynthetic aggregates of their own type. In the course of the growths and multiplications of these units, some of the photosynthetic ones either died or came into contact with the advanced non-photosynthetic ones; in both cases their contents became food for the latter group.

If we assume that the aggregates which developed the primitive photosynthetic processes were members of the "D-pentose-L-alpha amino acid group" (an assumption which is fully supported by the structural makeup of the contemporary organisms), then we can explain why the members of their opposite group lost their battle for survival. The aggregates belonging to the "L-pentose-D-alpha amino acid group" did not produce effective photosynthesizing mechanisms in time to meet the crisis of the shortage of supplies, and the members of this group could not utilize the products of the photosynthetic aggregates belonging to the "D-pentose-L-alpha-amino acid group"; thus they must have been slowly starved out of existence.

Establishment of General Patterns of Life

The development of the process of photosynthesis ended the era of precarious existence for the evolving aggregates. Life had become established on Earth. From then on it was only a question of improving the efficiencies of the systems, with the stimulus for improvements coming from the competition among the evolving units.

In due course, two main groups of aggregates with the "D-pentose-L-alpha-amino acids" composition gained dominance over the rest. These were: (1) the photosynthetic ones that could efficiently utilize the energy of the Sun and synthesize the vital compounds; and (2) those that could use the solar energy indirectly by first breaking down the photosynthetic and the weaker non-photosynthetic aggregates through the use of primitive yet efficient enzyme systems and then consuming the resulting compounds. The first group were the forebearers of the plant kingdom and the second group constituted the ancestors of the animal kingdom.

The Third Evolutionary Stage — Progress Toward Stability

As evolution progressed, the aggregates became more complex in composition, moving to what may be called the "coacervate stage"—that is, they were composed of primitive protoplasmic material covered with a simple protective layer. As usual, the improvement came by the process of trial and error. The protoplasmic material carried all the metabolic and reproductive functions. The primitive membrane consisted of a net-like cover partially impervious to the action of most hydrolytic enzymes from outside; it prevented the foreign enzymes from entering but at the same time permitted the smaller molecules to pass through.

In time the coacervates evolved into cells; this transformation probably took about as much time as it took for the coacervates to be formed from the prebiotic compounds. We know so little about this period that it would be virtually impossible even to conjecture about the biochemical evolution that occurred but attempts can be made at following the changes that took place in the morphological structures. Most probably, the evolution of coacervates into cells passed through two principal stages, the formation of organelle-type organisms, and the evolution of these units into different types of primitive cells.

Evolution of Coacervates to Cells

It is not difficult to visualize how the coacervates evolved into organelle-type organisms; the change was primarily a question of improving the efficiency of the existing mechanisms. The stimulation for the new advancements came from the competition among the units and the need for adapting to the ever-changing environmental conditions. It should, however, be mentioned that the main difference between the contemporay organelles and the most advanced organelle-type coavervates of ancient times lies in the degree of their self-sufficiency. The organelle-types of the pre-cellular period were more or less independent entities; the present-day organelles, though having the capability to self-replicate, are entirely dependent on the metabolic activities of their hosts.

The organelle-type coacervates were most probably the central units from which the eukaryotes, the prokaryotes and viruses evolved. The diagram in Figure 55 shows the suggested pathways for these evolutions, with each pathway representing a general pattern for the evolution of a variety of organisms from different types of organelles. Since there was a comparatively large variety of organelles with their own specific lineages, each successful evolution resulted in a cell with its own specific genetic code. This means that *there never was an original cell from which all the living descended.* Instead, many types of cells evolved independently of each other, with each new cell establishing its own specific lineage. This is the only way one can explain the diversity in the structures of DNA

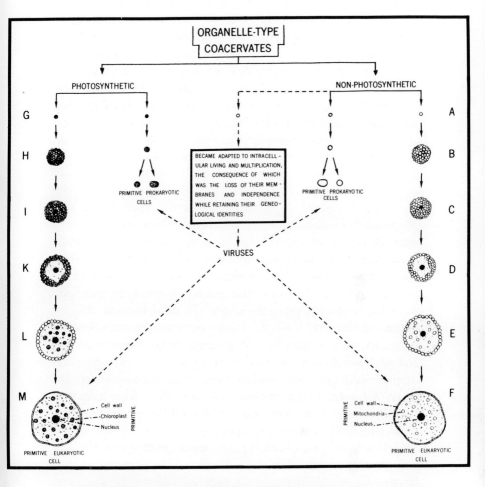

Figure 55. DIAGRAMMATIC REPRESENTATION OF THE EVOLUTION OF PHOTOSYNTHETIC AND NON-PHOTOSYNTHETIC EUKARYOTIC AND PROKARYOTIC CELLS FROM ORGANELLE-TYPE COACERVATES.

macromolecules among the genealogically unrelated biological species.

The evolution of the organelle-type coacervates to primitive blue green algae and bacteria was the simplest of all the transformations. Basically, it consisted of enlargement in size and the improvement in efficiency and complexity of the mechanisms. Natural selection then determined the types of organisms that could survive and proliferate.

The proposed pathways for the evolution of both the photosynthetic and non-photosynthetic coacervates to eukaryotic cells consist of the formation of colonies which then differentiate into nuclei, protoplasmic material, dependent organelles and cell membranes. According to this scheme, the descendants of a coacervate occasionally form a colony. Then infrequently a member of one of the colonies becomes enzymatically hyperactive and causes the dissolution of the membranes of the neighboring coacervates. The products of this breakdown then coalesce and hold the rest of the unaffected organelle-type units together. Under the direction of the primitive nucleus, division follows division. As a result, the components gradually become integrated, and eventually each assumes a specific function. The coalesced contents turn into a protoplasm-type material; the outside units gradually join to form a protective cover; and some of the internal organelle-type members which have succeeded in resisting the excessive activities of the central body differentiate into specialty components of the evolving cell, such as mitochondria-type and chloroplast-type organelles.

Evolution of Viruses

A minor evolutionary pathway in the diagram relates to the way a small number of coacervates were transformed to viruses. As is to be expected, during the cellular evolution most of the organelle-type coacervates that could not keep up with the general advance gradually perished. But some of them managed to survive by becoming adapted to the evolving cells; they learned to live and multiply inside their host.

The principal of adaptation was the same as that of the mitochondria and chloroplasts, but with a notable difference: the organelles capable of becoming adapted possessed the same genealogical makeup as their host cells, but the DNA structures of the organelles destined to become viruses differed from those of the invaded cells. This type of adaptation required the loss of the protective covers of the intruders so that they could effectively benefit from the metabolic activities of their hosts. The only part of the original unit that remained intact was the sequences of the nitrogenous bases in their DNA (sometimes RNA) strands and the structures of their related proteins. In many cases the cell managed to eliminate the invader through its enzymatic activities, but those intruders whose molecular configurations could resist the actions of the cellular enzymes escaped destruction. There was a continuous struggle between the invader and the invaded; eventually the parasite became adapted to live with its host, to draw as much benefit as it could, and to damage the cells as little as possible.

This more or less completes the discussion on the origin of life. Since very little can be added to the wealth of available knowledge regarding the "Darwinian evolution," we shall bypass the field of post-cellular evolution. What remains to be completed are a few statements concerning certain aspects of the evolution of life and some philosophical opinions regarding the possible role of Man in the general evolutionary schemes.

Comments

One of the arguments in favor of the hypothesis that depicts a very gradual and lengthy series of developments during the origin of life is the total dependence of life on extremely complicated and highly integrated metabolic and reproductive processes. Such a delicately balanced combination of systems and mechanisms even in the simplest of living organisms could not have come into existence within a relatively short time through certain coincidental physical and chemical events. The Darwinian evolution, a continuation and replica of what occurred during the pre-cellular era, also supports the concept of gradual and methodical evolution of prebiotic compounds to organized cells. In this respect the post-cellular evolution was more or less replicating the patterns of pre-cellular evolution.

All of the known principles relating to the evolution of life, such as natural selection, survival of the fittest, adaptation, etc., are based on observations of the natural outcome of the processes of self-replication. This continual and blind drive—directed by the DNA macromolecules to reproduce more and more of the members of the same species on the surface of a planet with a restricted space, limited material supply, a fixed rate of intake of energy and varying environmental conditions—has resulted in the development of a number of interdependent phenomena which have induced the formulation of these basic rules about the evolution of life.

Among the fundamentals related to the evolution of life, a single fact stands out above all others: the struggles for survival are merely endeavors of the living units to partake in the movement of the limited amount of energy flowing in and out of the Earth. In this regard, Man's attempts to become and remain supreme among the living are merely efforts on his part to become an ever larger link in this energy-transferring chain.

The Role of Man

In assessing Man's potential influence on the universe, the first and most important fact to be remembered is that his dominance over other units of life is not self-made, but is a natural consequence of the evolution of life. He has become what he is, not because of his own volition but because the trend of life's evolution has forced him into that situation. The significance of this statement is that Man, with all his cultural and scientific achievements, his capability of foreseeing and preparing for the future, and his power to devise methods for using the natural resources for his own benefits, can never become independent of the evolutionary system that raised him to his present position. Whatever he does or will do, even what we may consider as meddling with the evolution itself, all will be absorbed by the trend. He will never be able to change the evolutionary trend of life according to his own desires, for one cannot change a flexible trend whose development and maintenance are dependent on the contributions of its components. This does not mean that Man has no great influence on the direction of the progress of the evoultion. He certainly has left his marks and will do so in the future. But that is as to be expected, for in every evolutionary system all the components continually contribute their share to sustain the trend.

The second fact about the role of Man is that, although he has managed to become a dominant force on Earth, his role in the overall evolutionary course of the universe is very limited, if not infinitesimal, to the point of total insignificance. His physical and physiological makeup and his total dependence on well-balanced and mild environmental conditions confine the limits of his influences to a relatively small number of planets in the Galaxy. As a result, it is virtually impossible for him to make any significant contribution to the master evolutionary trend, the evolution of matter.

Whatever the history of early Man, it was his attempts at dominance over the members of the animal kingdom that brought out his power of reasoning; but the real advancement came when men sought dominance over each other. The resulting interspecies conflicts had great influence on accelerating the brain's evolutionary progress. Inventions had to compete with other inventions, new tools with newer tools, original weapons with more original weapons, innovative tactics with more innovative tactics and cunning with more effective craftiness. Battles and wars, therefore, were the basic ingredients for the mental progress of Man and eventually his civilization. Civilization, however, has not brought him peace and at this time the prospects are for more struggles. This raises some doubts about his retaining his superior position in the evolution of life. The excessive zeal of some members of his species to dominate the rest, a heritage passed down from generation to generation, may bring about the termination of his reign. Still, the possibility of his relinquishing his supremacy seems to be extremely small. He has a vast amount of knowledge, he is creative, he has become adapted to all types of environments and to living all over the Earth, and above all, he has a very deep sense of species preservation, which makes him watchful of committing acts that would result in his total extinction.

Henceforth two alternative courses are open to the evolution of Man. He can take part in intensive competitions for dominance and follow the pattern that has been in existence since he gained consciousness. This course could end in the release of the tremendously destructive forces mustered by scientists. But Man can use reasoning, the other aspect of the evolution of his brain, to suppress the desire for supremacy over others. The first course will undoubtedly hasten the approach of his destruction and dissolution; in this case all that will happen will be the loss of Man's position in the evolutionary trend. One should always remember that it is only to mankind that mankind is important and that there are no feelings, purposes, goals or urgencies in life's evolutionary trend. Therefore, should Man disappear from the scene, the evolution of what is left of life will proceed in its own plodding way and the human failure would be merely another one of the failures that happen regularly in all evolutions. But if reasoning is to prevail, and the indications are that it may, then the present stage of Man's evolution will be prolonged as long as the cosmological events will permit. In that case we can expect that, in due course, Man's restless mind and his adventurous spirit will direct some members of his species to spearhead the overcoming of the confinement of the Earth and the solar system. For this he will require vast amounts of controllable energy.

Since his emergence, Man has had to manipulate the natural-energy balances and redirect the flows of energies to the channels of his own choosing. In the first stages of his evolution this interference was negligible; but as his knowledge grew and his population increased, a variety of methods for extracting energy from all possible sources was devised. With the advent of modern science and technology, the use of energy has sky-rocketed to such levels that, at the present rate of expenditure, almost all the important chemical fuels will be exhausted in less than a relatively short period of a thousand years. After that, reliance has to be placed on the direct or indirect utilization of energy from the Sun and whatever can be developed in the line of conversion of matter into energy. This trend is obviously a part of the evolution of life and makes one ask two pertinent and inter-related questions:

1. Are all the activities related to the consumption of energy a preview of Man's preparation for the usage of much greater

amounts of energy from the more exotic sources?

2. If that be the case, is Man then preparing himself to enter the main trend of evolution of matter by learning the methods for the total conversion of matter into energy, thus replicating the evolutionary pattern of stars?

To answer these questions we have to take another look at Man's characteristics and the direction of his progress.

We are now at the dawn of space travel. Some men have already reached the Moon and there are other adventurous people who are willing to take their chances at landing on the neighboring planets, without any urgent reasons for doing so. In other words, whether we like it or not the trend is toward the exit of some members of the species from the Earth and their entrance into the space beyond the solar system. What has been done so far, then, may be considered as skirmishes in an expanding and inviting field—but these will be of no avail unless Man comes into the possession of truly powerful means of propulsion that would propel his ships in space at close to relativistic velocities. The only source for such tremendous amounts of energy is from *antimatter* material.

The most obvious sources of *antimatter* material are the comets; they are relatively small, compact and dense in structure, will not vaporize easily and are frequent visitors to the solar system. This may sound like an idea taken from the books of fantasy, but the truth is that with enough ingenuity and close cooperation between the technologically oriented nations, these objects in the skies can be captured and made to orbit the Earth, or preferably the Moon, at safe distances. The technique for the capture of these objects would be based on the principle of making a comet generate excessive energy at a chosen area on its surface—for example, in front, if it is desired to reduce its velocity, or on a specific side if a change in direction is sought. Since the energy produced by a comet is the product of its continuous encounters with $+matter$ atoms, then in order to increase the production of energy at an indicated point on its surface one would have to force the encounter of the selected area with high concentrations of $+matter$ gases. It is obvious that this is not an easy task and cannot be achieved with our present-day techniques, but we do have the capability of developing the technology within less than a century.

As compared to the difficulties of capturing a comet, the development of the technology to use the captured *antimatter* masses as sources of energy, either for space travel or for industrial purposes, seems an improbable if not an impossible project. However, if we look at it carefully we find that we are already in possession of two pieces of vital information for entering the field. The first one is the formation of the radiation zone around the *antimatter* mass when surrounded by +*matter* gases. This keeps the two antagonistic masses at a distance and thus provides conditions for keeping the reactions under control. The second one is the possibility of existence of *antimatter* either inside or in the upper atmosphere of Jupiter. The red spot might have been caused by a comet embedded in the liquid hydrogen on the surface of the planet, and the possible presence of *antimatter* gases in the upper atmosphere of the planet could provide us with clues for controlled +*matter-antimatter reactions*. The successful utilization of the *antimatter* energy, therefore, may not be as hopeless as it seems at first glance. Although the project belongs to the future, it cannot be considered out of reach.

The discussion about the use of *antimatter* as possible sources of energy points to the coming direction of life's evolutionary trend. Life originated principally because the products of the annihilation of matter encountered the proper conditions on Earth and caused the formation of a system that was and still is totally dependent on the incoming energy created by the +*matter-antimatter reactions*. The system in turn produced a thinking being with the capability of understanding the nature of matter. This faculty is certain to direct him to attempt to bring under his control a very small portion of the very same types of reactions that brought about his origin and his evolution. This means that the function of Man in the main evolutionary trend will be that of space scavengers. By removing the small cosmological debris and converting them to energy he will be contributing a very minute share to the general trend of evolution of matter.

The three and one-half billion years of the post-cellular evolution have thus culminated in the emergence of a species destined to replicate the general evolutionary trend in the universe. In this capacity all the members of the species are guided by the master trend, the evolution of matter. As such they are servants to this unthinking, merciless and indifferent system whose power comes from the continual thrust of innumerable universal units toward the establishment of total uniformity of motions.

How far Man can go to help in this movement is anybody's guess, but we can be sure that whatever he is doing or attempting—his loves and hates, his successes and failures, his wars and peace, and above all his sciences, cultures and industries—all are aimed at preparing him for playing the role directed by the cosmological forces. To our way of thinking, this probable fate may be considered as ridiculous, absurd, exciting or any other adjective we may choose for the description; but before we pass judgment we should take note of the indisputable fact that the evolution of matter is totally oblivious to the feelings of men, their expectations, and their arbitrary codes and standards.

INDEX

THE AUTHOR

Bahman K. Shahrokh was born in Teheran, Iran, in 1913. He received his Ph.D. (1943) in microbiology from the University of California, Berkeley. His lifelong interest in the problems of various evolutionary systems has resulted in this book, a summation of his theories regarding the origins and evolutions of matter and life in the universe.